Tim Cole ♦ Ossi Urchs

Digital Enlightenment Now!

How the Internet is making us better and smarter and in the process changing just about everything around us!

Forsthaus

Imprint

The text of this work is licensed under the Creative Commons Attribution 4.0 International License. To view a copy of this license, visit *http://creativecommons.org/licenses/by/4.0/*

The authors feel that it is a waste of time to try and protect ourselves from illicit copying. It is in the interest of readers, authors and society that cultural knowledge flow as freely as possible.

However, under the Creative Commons license we reserve the right to commercialize content we have created by offering this book for sale in various forms.

Herstellung und Verlag:
BoD - Books on Demand, Norderstedt
ISBN 9783734768385

© Forsthaus-Verlag, Forsthausgasse 80, A-5582 St. Michael im Lungau, Tel. +43 (6477) 20253, *forsthaus-verlag@gmx.at*

Content

Introduction: Living at Internet speed 7

1. Why We Need Digital Enlightment 19
   What the teapot teaches us 20
   Understanding the new 23
   10 Theses on Digital Enlightenment 25
   Categories for a New Enlightenment 39

2. The History and Future of Networking 43
   Networking without networks 45
   Metcalf's Legacy 52
   Life in the swarm 56

3. Thinking in Real Time 65
   Digital Natives are not a new generation 70
   Digitalization changes our perceptions 73
   Why "multitasking trauma" is just a myth 77

4. The Networked Human 85
   Game nuts are really quite peaceful 87
   Evolution in fast-forward 89
   The sixth Kondratieff 92
   Making a map of the brain 94
   In praise of distraction 96
   Neuphobes and neophiles 100
   Everybody's gone surfin' 104

5. Generation Now! 111
   Digital bottle babies 113
   Welcome to the Facebook Society 115
   It's all mine, mine, mine! 119
   Time killers or time savers? 121
   No life without Facebook 125
   Everybody speaks for the company 129

6. The New Life Plan 135
   After work is so yesterday! 136
   Work without borders 138
   Digitalization versus industrialization 145
   Digital Bedouin seeks digital oasis 154

7. Welcome to the Global Village! 161
   Pulcinell's Secret and the invention of privacy 162
   What happened in the barn 163
   Going public 164
   Digital Omerta 168
   The Rumpelstiltskin effect 172
   Under the digital veil 174
   Agents, avatars and anonymity 176
   The right to remain anonymous 179

8 Information Wants To Be Free 183
   Black holes in cyberspace 184
   The new sense of justice 187
   A recipe for pirating 189
   Art without copyright – copyright is not an art 190
   Is intellectual property theft? 192

Old content, new context 196
Information without context 199
Bloggers –amateurs take over the newsbeat 202
The twilight of journalism 205

9. The Great Earthquake of New York 211
The fear of freedom 212
Pity the messenger 214
The legacy of the Twin Towers 216
Watch the watchers! 217

10. The End of Utopia 225
Politics in real time 227
Digital particularism 234
Doctrines are for dummies 238
The future off intelligence 245

11. Think For Yourself! 251
A new vocabulary for a new ethics 252
All will be made clear 254
The future is open 256
Autonomy as a system 259

Afterword: How we wrote this book 263

Introduction

Living at Internet speed

In "Through the Looking Glass", Lewis Carrol's charming little book for children of all ages, the Red Queen takes Alice by the hand and pulls her away, running hard until the little girl is tired and has to stop to catch her breath. To her surprise, she discovers that she is still standing right where she started off. When she complains, the Red Queen replies: "You must come from a slow sort of country! Now, here, you see, it takes all the running you can do, to keep in the same place. If you want to get somewhere else, you must run at least twice as fast as that!"

Welcome to the land behind the computer screen where we all feel that things are happening much faster than in the real world; so fast we have problems even understanding what is going on. Everything, it seems, is happening at fast forward.

The Internet, we all know, has given us a new way of reckoning time: "Internet Speed", a time frame some used to define as seven to eight times faster than real life. Which calls "dog's years" to mind which most of us believe are seven times shorter than a human year.

This is hard for many of us to grasp, and not only the elderly among us. It has more to do with adaptability, flexibility, open-mindedness and willingness to accept change than with mere physical age.

Twitter is a good example. People seem to be divided into two camps when it comes to these tiny snippets of text which someone once described as "Internet telegrams". On the one hand there are those who think Twitter is just about the stupidest invention ever; the others believe it's the greatest thing since the telephone.

Twitter allows us to send messages consisting of 140 characters each directly from our PCs or smartphones. You can read these digital memos on Twitter's website, but since tens of thousands of "tweets" are being sent out every minute no one can hope to keep up. So instead we all get to choose a number of Twitter authors whom we want to "follow". This means that we only see tweets from people we want to keep an eye on.

You would be surprised how profound and intelligent a message can be that is only 140 characters long! But of course, composing one of these little gems is really hard work, so most of the messages we see on Twitter are boring at best, inane at worst. But very occasionally one will turn up where we say "wow!"

99 percent of all tweets may be pure drivel, but if you catch the single inspired one in a hundred, it's well worth your time and trouble. At least that's what fans of Twitter say, and an astonishing number of them

are mature adults; members of the "baby boomer" generation rapidly approaching retirement age or already passed their due date.

One of our acquaintances who goes under the Twitter name "@lusches" is in fact the middle-aged owner of a butcher shop who fires off dozens of tweets a day. He describes making his latest batch of bratwurst, tells us how delicious his lamb chops taste or what he is preparing for his next catering job. You would think this guy has his hands full, but no: somehow he always finds the time to tweet. "It's become second nature", he tells us.

I am me and I am here!

Twitter is a heady mixture of self-expression and instant communication. Many of us find this fascinating. "I am me and I am here": that's the core message constantly going out through this new conduit, the quintessential "new medium". In fact, of course, it is just one of thousands of technological innovations that are bringing us closer and closer together and that at the same time are changing our perceptions of reality.

Of course keeping abreast of digital innovation puts pressure on us as individuals. Every attempt to communicate, every e-mail, every short message and every tweet feel like a cry for help: There's someone out there who desperately wants to get in touch, so please pick up, please send something back, talk to me! And since we are who we are we keep trying to respond to every attempt at establishing contact.

"I'm in a meeting", someone said recently, but he then went on talking on his mobile for five minutes before hunging up. Heaven knows what the other people in the room were doing in the meanwhile; possibly checking mail or texting.

We all feel it in our bones: our lives are speeding up! We sit chained to our benches rowing to the beat of a drum that is going faster and faster and we can't even see the drummer who is driving us forward. We are all little Alices, running as fast as we can hoping that we can stay where we are and not fall behind.

The author Ray Kurzweil calls this "accelerated progress", and he believes we these are still at the very beginning. "The rate of progress of an evolutionary process increases exponentially over time", he writes in his book "The Age of Spiritual Machines"[1]. If this is true it means that technological progress in the twenty-first century will be equivalent to the progress made by mankind during the last 200 centuries.

But we have a slight problem here: Humans live their lives linearly. Besides, we all know that exponential models have a tendency to collapse. They're all just Ponzi schemes anyway, aren't they? Sometimes the curves on the graphs just keep going up and up as if the laws of gravity have been suspended, but of course they aren't really. Just take the "Dotcom Bubble" that burst in 2001, bringing all our dreams of instant wealth crashing down around our heads, dis-

---

[1] Kurzweil, Ray, The Age of Spiritual Machines (Penguin Books) 2000

appearing down the rabbit hole just like little Alice in "Wonderland".

Which brings us, in a roundabout way, back to the story of Alice in the land behind the looking glass and the question of how fast we need to run if we don't want to fall hopelessly behind. Never has Lewis Carrol's cute little story seemed as crucial to understanding the world as it is today; at time when social and psychological change is happening at Internet speed.

Digitalization moves markets

Consider the single biggest "megatrend" of the past two decades, namely digitalization. Everything that can be digitized will be digitized, as we will explain later in this book. The reason is simple: digitalization pushes costs down. So digitalization is rapidly becoming a huge market factor; one that directly impacts both our economies and our lives.

All you need to do is compare pre-digital products with their digital successors and see how prices for digital merchandise have fallen. A couple of decades ago a vinyl long-playing record cost about $20. Twenty years later, a "semi-digital" CD still costs a pretty packet, possibly $15. But if you are actually willing to pay for digital music you can download the same content directly from a website and it will only cost you a few cents, if anything at all.

Digital products are cheaper because the variable costs of distribution are potentially zero. This affects not only packaging and logistics, but also warehousing and salesrooms, just to name a few of the more important economic factors. That's why Amazon can challenge classical "stationary" retailers today not only in the digital marketspace, but with analog products, too. After all, their entire business model is based on digitalization.

The price decay brought about by digitalization inevitably leads to more and harsher competition. The only way vendors (including online shops, but "old-world" manufacturers as well) can hope to cope is by establishing more and more direct contact with their customers, cutting out the middlemen through the increasing use of network technology.

Digitalization doesn't just mean price degradation: it also causes breathtaking acceleration in all areas of technology, including the media. From Gutenberg's "invention" of the printing press (actually, the Chinese were using movable type a thousand years ago) to the rise of mass-circulation newspapers in the early 19th century at least 350 years had to pass. Tim Berners-Lee's idea of a "World Wide Web" took only five years from its first inception at the CERN laboratory in Geneva to becoming a revolutionary force.

The concept developed by "TBL" (as Internet aficionados call him) was originally meant to connect scientists at CERN and the results of their research with their peers around the world. It was based on long-established technologies such as the "HyperCard" software that enabled computers to create digital versions of simple, everyday filing cards that could be connected to each other by "links".

The stroke of genius here is that on the Web, cards stored on different computers in different parts of the world can be linked together too. In fact, the computers at CERN were already connected through the so-called TCP/IP protocol. All TBL had to do was write a new software "stack" to stick on top of the existing protocols. Thus, "HTML" was born: the Hypertext Markup Language which is still spoken by every Internet device today. And this only took him a couple of days.

The dramatic change brought about by the development of the Web and other, similar software projects was based on the principle of open and easily accessible standards. This strategy of open standard development not only sped up software development but also the developer a big boost in efficiency. By contrast, traditional media are closed and restricted both in quantative term (the speed in which they can be developed and deployed) and qualitatively (the strategy's "openness", e.g. its ability to rapidly integrate new technologies and applications).

This is exactly the strategy Tim Berners-Lee chose when he combined the open standards and protocols developed for the Internet with his own unique contributions. His brilliant idea was to use HTTP ("Hyper Text Transfer Protocol") to allow every single user to query a (web) server which in turn would send back an "answer" in the form of digital data using the same protocol. This changed the world.

To display these data packets on the client's computer he used standards called "markup languages" which he developed further into HTML. Thus he was able to present a document or image as a digital "page", a metaphor borrowed from an earlier age when books and magazines still had paper pages.

This allowed TBL to develop all the important elements of his system for transmitting and displaying digital information through the Web in record time. And because it was based on open standards, other scientists and software programmers could improve and expand the system any way they wanted.

Today, of course, we send and receive not only documents and images, but voice and video files, and we do it in "real time". Developers can design new ways of displaying data, technicians and programmers can dream up new functions to make Web content more useful or more fun for the user. The Web has proven to be endlessly adaptable and flexible – and therefore superior to any other mass medium in history.

The acceleration caused by digitalization and worldwide networks have produced (or at least fostered) so-called "disruptive" developments in technology; advances that are literally able to dislocate or call in question entire industries and giant corporations as well as their business models.

Take the manufacturers of navigation devices, for example. A few years ago these companies lived in a cozy little niche where they produced relatively simple gadgets, stuffed them with sophisticated software and sold them at a fantastic markup to a market that seemed to be growing steadily. Along came Google with its own maps and software for smartphones and tablets using their proprietary "Android" operating system, and overnight the flourishing navigation device business was almost wiped out, leading Google's CEO Eric Schmidt to famously describe navigation aids as a "zero-billion dollar business".

Networking always leads to change

The second gigantic trend today is directly connected with digitalization. Networking on a global scale is changing the world around us. It turns out that change is inevitable in a networked world. It not only affects the systems that are being connected, but the business processes that are run through them as well, along with the people who use them to work or communicate. Everything has to follow the rules of Digital Transformation, about which much more later.

Just as digitalization leads to acceleration, networking leads to change. Both are simply parts of their natures. A good example was provided by Vinton Cerf, the legendary "father of the Internet" and inventor of the TCP/IP protocol which is used all over the Internet today. Vinton was our guest in a TV talk show we both hosted for the German *cnn* affiliate, *n-tv*. What would happen, Vinton asked the live audience, if you were to connect an Internet icebox with a pair of Internet-enabled bathroom scales? We don't really know, he said – except that something will change. Maybe we will come home and find our refrigerator full of diet food, or maybe we won't be able to open the icebox door. Or maybe something completely different will happen, but we can be pretty sure something will.

Networking is changing the way companies do business. However, it isn't always immediately apparent where this change is going on, what the result will be and how it affects the bottom line. The greatest challenge managers face in a digitalized and networked business world is discovering exactly how these trends are impacting their enterprise and its business environment, and then reacting quickly and intelligently. Those who are best at adapting to Digital Transformation will be the winners, and the race is to the swift.

Digitalization and networking thus prove to be two sides of the same coin; complementary forces that operate at Internet speed. Not only is the development of technology touched by this, but our entire way of life along with our businesses and our personal wellbeing. Welcome to the Digital World!

Chapter 1

Why We Need Digital Enlightment

Why We Need Digital Enlightment

"The incredible expansion of knowledge in our times and the rise of new sciences make it hard for us to discover the truth and put it to use".

Lawrence Durell: Justine, 1957

It has become fashionable to blame the Internet and digital networks for all the ills that face mankind. Cultural pessimists are having a heyday, like the American "computer-artist" Jaron Lanier who gloomily sums up 20-odd years of Internet development with the words: "how we have screwed things up". In Germany Frank Schirrmacher, the publisher of the prestigious newspaper "Frankfurter Allgemeine Zeitung", declared a "cognitive crisis" shortly before he died in 2014. The psychologist and bestselling author Manfred Spitzer diagnosed a kind of collective "digital dementia" among heavy Internet users and believes that "the Internet is making us dumb", a sentiment echoed by Harvard Business Review writer Nick Carr who asks, "Is Google making us stupid?"

Cultural pessimists have always been around. Socrates, Plato wrote, bemoaned the introduction of writing which ancient Greeks brought in from Egypt, since it meant that people no longer had to memorize long texts. He worried that this would "plant oblivion in our souls". In 17th century Britain, the philosopher Robert Burton, author of *The Anatomy of Melancholy*, complained about the flood of books caused by the recent invention of the printing press; a kind an analog version of "information overload" which cultural pessimists complain about today.

Most of these naysayers see the Information Society strictly in simple mechanical terms. All you have to do, they say, is inundate people with information and pretty soon they'll stop thinking for themselves, stumbling instead through life as though under drugs or remote control. These "critics" of the Internet seem unable to believe that human beings already have (or are in the process of developing) the ability to separate clearly between relevant and irrelevant information. For them, mankind are like cattle contentedly chewing their cuds of information, prodded along by media-savvy herdsman like themselves towards an uncertain future which is beyond their power to control or shape.

What the teapot teaches us

One of the authors of this book once had the honor of being invited into the former home of Konosuke Matsushita, the deceased founder of Panasonic. The villa, which is now a private museum, is located on the outskirts of Osaka, where the Matsushita Corporation has its headquarters. There he attended a traditional tea ceremony during which a tiny lady in a formal green kimono poured tea into his cup with an incomparable grace of movement. Intrigued, he later went out and bought just the kind of high-spouted Japanese teapot she used, but back home, whenever he tried to pour from it, the tea spilled out and formed a puddle on the table. Obviously, he lacked what the Japanese call "wa"; a kind of inner peace and balance that can inhabit not only people but objects and even places, such as the villa of Konosuke Matsushita.

One day, though, the author found that somehow he had mysteriously and unconsciously mastered the art of expertly pouring tea from a Japanese teapot. At least he no longer spilled most of it. Which leads to a fascinating question: Had he, the human being, learned over time to pour, or had the teapot in fact taught him how?

This is hardly a trivial question in an age where "digital overload" and rewired brains are a popular topic of discussion. The really important question is about human self-empowerment: Are we the masters of our computers or their slaves? Does Google tell us what to think? Is multitasking part of mankind's adjustment to a rapidly changing communications environment, or can it be described as a form of bodily injury being inflicted on am unwary populace, as Frank Schirrmacher once wrote? Let's put it another way: Are we being driven, or are we the drivers; free agents and members of a species that is especially adept at modifying itself as a way to adapt to changes in the surroundings?

Norbert Bolz, a good friend of both of us and a professor at the Berlin Technical University, who iswidely considered the leading postmodern philosopher of media science in Germany, gave a convincing answer which he recently posted on the blog *czyslansky.net*. It goes like this:

> "More and more of us despair of being able to manage their own attention spans. Actually, it's all about a quite simple question namely: What is really important? In order to reach an answer to this question we need to reduce complexity. In our search for orientation media, technologies act as filters, but in the end it is all about human judgment. Does this mean that digital reading makes our brains turn digital? Behind that kind of question lurks the latest form of cultural pessimism."

This wholesale rejection of things new like the Internet or digitalization isn't especially unique in the history of thinking about thinking, and it isn't even very original, as the examples we just quoted show. Still, this brand of technical pessimism does manage to ring a bell again and again for a large number of people, especially in times of fundamental social change.

This book does not ask whether the human brain operates "digitally" or not, although an increasing number of cognitive scientists and neurologists do believe that this is so. Instead, we propose to refute the generations of cultural pessimists, from Socrates to Schirrmacher, and in the process contribute to a necessary discussion about what we think of as "Digital

Enlightenment"; namely a new intellectual effort to describe the basic requirements for transforming humans and humanity into what we will call the "digital society".

Understanding the new

The great American thinker and author Robert Anton Wilson divided humans into "neophobes" – people for whom the new and unknown is a cause for angst and doubt – and "neophiles": people who are eager at all times to explore the new (a fancy way of describing nerds).

While this method of defining mankind may be both helpful and amusing, it doesn't really increase our understanding of how digitalization and digital networks are causing radical change, much less whether these forces are either productive or useful. Our aim in this book is to show that the "signs of the times" are truly novel and unique, and that instead of assigning values based on old and tired concepts we need to develop a new way of thinking.

Not that we aren't all constantly modifying our brains as we go along; something which worries the cultural pessimists no end. Wilson calls this "metaprogramming the human bio-computer".

"Every time we learn a new fact or skill we are actually rewiring our brains", writes Steven Pinker, a famous Harvard psychiatry professor and bestselling author, in an essay for the New York Times. But, he continues, "neuronal plasticity does not mean that our brains are lumps of clay that are being slapped into new shapes by every new experience we have." In other words: Experience doesn't reduce our brain's ability to process information. Quite the opposite: it enhances that capability!

Which brings us back to the question of digital overkload. The notion that our brains deteriorate or are being crippled by exposure to too much information and communication is essentially misanthropic. Our brains, like the rest of us, don't regress, they adapt, adding new capabilities through evolutionary adaptation. This has been going on for as long a mankind has existed and will continue for as long as we humans are around.

The story of the teapot illustrates this very neatly. Of course the teapot had an effect on its "user". It taught him a new skill, presumably by improving his hand-eye coordination, and it helped him to pour from it with the necessary "wa". In the process, hundreds or possible thousands of synapses in his brain were rewired or reprogrammed. So yes, he was the object of an invasion of his personality. But he was definitely not the unwitting victim; in fact he was an active participant in the process.

With this book and the theses it contains we hope to initiate a new school of thought and kick start a social discussion in the spirit of the classical European Enlightenment of the 18th and 19th centuries. Our aim is to ensure that we will collectively be able to approach the future with a new mindset, in the process discovering how we can put the new to its best possible use, or else determining that it deserves to be thrown onto the trash heap of history instead.

10 Theses on Digital Enlightenment

> ***Thesis 1: Everything that can be digitalized will be digitalized. Everything that can be connected will be. And that changes everything!***

The massive trend towards digitalization, as shown in our introduction, has economic root that reach back to something called, rather confusingly, "Moore's Law". What Gordon E. Moore, one of the founders of Intel Corporation, actually described back in 1965 was less a law than a hypothesis, albeit one that has remained valid to this day.

Moore's "law" simply describes the tendency of digital gadgets to double their capacity approximately every two years. This ability to sustain growth that is exponential (a phenomenon that we will hear more about later) also leads to the halving of costs for digital computing power approximately every two years. More than any other factor such as the elimination of variable costs, this causes priced for digital products to decline rapidly. This fall not only effects microprocessors and digital storage media, but even everyday devices such as refrigerators and washing machines,

TV sets and telephones, since they are all loaded with microchips nowadays. Indirectly this also influences the logistical distribution systems and other services these markets depend upon, since they too are increasingly being managed digitally.

As always when markets suffer a massive fall in prices vendors try to find ways to "cut out the middlemen" by reaching out directly to their potential customers. Global connectivity is the perfect way for them to interact with consumers around the world, so more and more manufacturers, dealers, markets and customers are "going digital" to a degree no one could have imagined in their wildest dreams just a few years back. Together, the ongoing digitalization of the economy and the global networks connecting markets and buyers have fundamentally changed the way the markets – and societies – work.

Thesis 2: Digitalization and networking are not like the common cold – they will never go away!

Digitalization and networking are radically changing the way we live and work, how we educate and entertain ourselves, how we makes purchases and do business, and most especially how we communicate with others. Literally every facet of human existence and most human lives are undergoing fundamental upheaval. We therefore must expect this development to continue far into the foreseeable future.

> **Thesis 3: The digital world is increasingly invading and becoming part of the real world. As a result, both are changing at breathtaking speed and at an unprecedented rate.**

Developments in technology and business are forcing change on society and on our personal lives, and we can't expect things to ever return to "normal" again. After all, nobody can rewrite history. We are currently experiencing a totally new and remarkable phenomenon, namely the coming together of the digital and the physical worlds which used to be strictly separated. In fact, it is getting more and more difficult to tell the two apart.

Navigation aids were once simply used to show us the way from A to B. As the digital world encroaches on the physical, we now expect our gadgets to use digital information to show us the "right" way. This can be for instance the fastest, or possibly the most scenic route, depending on our individual circumstances. "Augmented Reality" does not simply mean "enriching". It changes our perception and our understanding of reality itself.

What we think of as "real" will increasingly reflect a mixture of digital and physical perceptions and experiences, displayed on smartphones or a tablet computers and enhanced by new, "wearable" devices such as Google's "Glass" or the Apple Watch. Thus we will become more and more accustomed to navigating something best described as the digital "infosphere" which will surround us just like the physical atmosphere of our home planet.

But don't worry: This won't turn us all into zombies marching powerlessly to the drumbeat of our digital masters. Quite the opposite, in fact. But we must develop our abilities to distinguish between relevant and irrelevant information, in effect filtering out the noise and tuning into those sources of information that will enrich our lives and help us become masters of our own destinies. And should some information turn out to distract us, then we need to learn how to be strong and smart enough to simply switch it off.

But being disconnecting from the digital infosphere will be like watching an old black and white movie today. Yes, we may even relish the experience as a form of ascetical and esthetical self-denial. By doing so, we will not only lose a (multi)medial dimension but possibly even heighten our concentration on other aspects of perceived reality. But we will always be aware that by simply pressing a button we can return to a richer and more satisfying dimension – one in which we will all feel more truly at home.

Thesis 4: Digitalization and networking create new technical, social, cultural and scientific conditions. In order to appreciate

> ***these changes fully we first need to categorize them before we can begin to fully describe and understand their new qualities.***

We are already at the point where our perceptions of reality have changed in many ways, as have the conditions under which we experience this new reality. Is a computer game less real than a romp through the woods? Is a love affair on Facebook less exciting than a flirt at the hotel bar? Perhaps the realization that digital experiences are based on binary code influences our assumptions; after all, if the only two possible digital conditions are "on" and "off", then maybe "real" and "unreal" are also just two sides of the same coin. In fact many of us have already learned that digital information can help us to understand the world around us in a more granular and modular way than ever before; without, it is to be hoped, losing sight of their basic unity in the sense of their inherent interconnectedness.

Digitalization and networking obviously influence the way we experience, understand and handle reality. But in fact the conditions under which we do this is undergoing continual change, too. The pace is so rapid and dynamic, our own experience of it so mind shaking, that we are in danger of forgetting to digest what is going around us on both intellectually and conceptually. The wealth of new developments and experiences in technology, society, culture or science demand and deserve a new way of describing them, a new terminology more suited to the Digital Age. Just as it is difficult to describe Quantum physics using the vocabulary of Newton mechanics, it will be impossible to define the digital world using concepts and phrases originally developed for the Analog Age.

> **Thesis 5: After 150 years mass media are losing their ability to forge communality and identity. As a result, communication will return to its roots as intrapersonal exchange, albeit through digital media.**

Human society is being reformatted as we speak. As our perception of reality shifts the familiar guard rails and signposts that normally help us find our way around are becoming irrelevant and obsolete. This affects not only religion and other belief systems, be they scientific or philosophical, but especially the old mass media and their ability to influence society, shape opinions and forge identity.

The "New York Times" may still claim to bring us "all the news that's fit to print", but actually those days are long gone. There are a number of reasons for this, both cultural and economic. The dramatic fall in the costs of disseminating information is causing mass media to lose their communal influence. At the same time, their traditional business models are crumbling, due largely to short-sightedness of the part of publishers and broadcasters who still fail to understand what has happened and who lack the vision (and the courage) to wrench the wheel around. The daily newspaper is not simply an endangered species; it is already dead to a generation of kids who are accustomed to get all the news they want free of charge from a wide range of sources: friends, bloggers, news show parodies such as Jon Stewart's "The Daily Show" or social media sites such as Facebook or Twitter.

In addition, audiences today are simply too media savvy to swallow the pseudo reality which mass media tries to create: Can anyone who has watched "Survivor" still believe in star worship? Mass media's market share among young consumers has fallen so low that producers and actors of such "media trash" can no longer generate the numbers they need to continue to operate – with the possible exception of a few overpaid football players and the occasional mass event such as the Olympics, the World Series or the Soccer World Cup. The days when people would put off calling home because it was time for the CBS Evening News are gone forever.

Mass media have had a good run. It lasted more than 150 years, but economically their business model is coming to an end. In their wake, they will leave behind the communication needs and habits of the "zoon politicon", Aristotle's "political animal"; a being whom it is impossible to conceive of in the singular. As a society we are being cast back on the original function of communication as an exchange; something mass media are inherently incapable of performing. As a result, we as a society are in the process of restructuring as we return to the Agora, the marketplace of ideas, only this time organized and transmitted digitally. As mass media fade away and die, we need to collectively develop the necessary mobility and strength of argument to exist in a digital society.

> **Thesis 6: Digitalization and networking inevitably lead to massive acceleration and disruption in technology and media. They are impossible to control; all we can do is hang on and enjoy the ride.**

Prices in the age of digitalization and networking are destined to fall because of innovation and market economics as well as other factors. Together, these will lead to dramatic acceleration in the areas of technology and media. It took approximately 350 years for Guttenberg's invention of the printing press to give us the first mass-circulation newspapers back in the early 19th century. It took the World Wide Web just five years to grow from the brainchild of Tim Berners-Lee into the largest "mass medium" the world has ever seen. And it took only three years for mobile devices to change the way consumers in the

U.S. and Europe, but especially in Asia and Latin America access and use the Internet.

Welcome to the Age of Acceleration, where technology and media evolve so fast that business models can't keep up and where established products and players can become obsolete almost overnight. The term "disruption" describes perfectly what happens if business models collide with rapid innovation and when rules change so fast that businesses can no longer keep up. In 1942, the Austria economist Josef Schumpeter[2] described what he called "creative destruction" as "a process of industrial mutation that incessantly revolutionizes the economic structure from within, incessantly destroying the old, incessantly creating the new." Nowhere has he been proven right more dramatically than in the Internet Age, where traditional markets suddenly collapse in a blaze of dynamic innovation in which rules are rewritten and established brands and corporate giants are forced to adapt – or die!

The iPhone is probably the best example to such a "disruptive technology". It allowed Apple to redefine the world market for mobile telephones and quickly pushed long-established leaders like Nokia and Research In Motion (makers of the Blackberry handsets) to the wall. In less than three years Apple catapulted itself from a newcomer to a dominant player and in the process "invented" a completely new mass

---

[2] Josef Schumpeter: Capitalism, Socialism and Democracy, Routledge; New Ed edition (1994)

market, namely the mobile Internet. Formerly the domain of a few techie geeks, online mobility has become part of everyday life, spawning an entire industry where suppliers, accessory vendors and service providers ply their respective trades. More importantly, the iPhone has spawned an entirely new lifestyle; one that has touched the lives and the self-esteem of millions around the world. The iPhone has been described as the "Swiss army knife of the digital age": an all-purpose tool for all things digital, and more importantly the "original" (something actually counterintuitive in the digital world) that others can only try to emulate.

However, it would be both presumptuous and naive to think we can accurately predict such disruptive markets (although this hardly keeps so-called futurologists and trend scouts from trying). Twenty years ago, anyone crazy enough to predict that texting would one day become the "killer application" of the worldwide mobile phone industry would have been laughed out of the conference room. And the same applies for those who believe they can "handle" disruption. The histories of enterprises such as IBM, Siemens, Nokia and Sony prove that only those who are able to reinvent themselves quickly when they run up against a disruptive situation can hope to survive: In the Age of Acceleration there is no such thing as a second chance.

> *Thesis 7: Life in such a fundamentally different world will become more complex, but not necessarily more complicated. In fact, it will probably be simpler and less stressful.*

For us users, disruptive moments in technology and media can actually be quite fun. Not, that we kid ourselves that we can manage them, but they can make our lives less complicated. If we choose, we can attempt to understand what's going on – but we don't have to.

Understanding where disruption is leading us won't be easy, but it may be well worth the effort. If we can come to grips with the changes happening around us, we may come to realize that life in such a fundamentally new reality is more complex, but not necessarily more complicated or more difficult. Quite the opposite: Those who understand how to make best use of disruptive innovation or are at least willing to try can lead richer and more rewarding lives. But first, we need to make an extra effort.

> *Thesis 8: Human beings, too, will need to consider the way they lead their lives. We need to develop the ability to think in a digitally and networked fashion – and in real time.*

As the German philosopher and cultural theorist Peter Sloterdijk writes in his book „*You Must Change Your Life*" [3] mankind in the 21st century faces the need to "think dangerously". Essentially, this will require quite a bit of strenuous mental exercise, but perhaps as we sweat on our mental treadmills, we can console ourselves with the thought that our object is to appreciate and penetrate the digital and connected systems which surround us; thereby in fact functioning the way the human brain is programmed to work, namely through synthesizing the many different perceptions our senses provide into an (interconnected) view of the world and of ourselves. Human thought, cognitive scientists now believe, works essentially like a computer network.

This does not however mean that the idea of man as a machine ("automaton"), a concept so popular with cultural pessimists in the past, is true. Instead, we need to understand the similarities between networked systems and human thought processes. This has less to do with the concept of "internalization" in the psychological sense and more with "differentiation" as understood by computer scientists – an essentially human form of information processing. The difficulty here lays not so much in the act of processing information but in the ability to cope with the results.

---

[3] Peter Sloterdijk: *You Must Change Your Life*, translation by Wieland Hoban, Cambridge, Polity Press (2013)

> ***Thesis 9: Concepts and truisms from the analog past are increasingly useless in guiding us to a digital future that is developing dynamically. Anyone espousing them will seem increasingly out of time or out of place; helpless, in fact.***

There is yet another difficulty we must master before we can seriously debate the future of mankind in the Age of the Internet: Are we culturally and linguistically prepared for what we are experiencing or not? We believe that notions and concepts dating back to the analog age are incapable of describing what is going on. They appear antiquarian, even quaint.

Harking back to the great media and technology debates of the 60ies and 70ies of the last century we quickly see just how outdated they are now. The hot-button issues back then such as commercial television or population census are about as relevant today as counting the number of angels dancing on the head of a pin. Even New Media are no longer new – or news. And the open-ended issues of data protection and privacy, which has been going through increasingly bizarre permutations especially in Europe, have a distinctly old-world flavor in the age of WikiLeaks and NSA. Listening to Jaron Lanier bemoaning what he insists on decrying as "digital Maoism" or when Frank Schirrmacher complains that his brain is being "mashed up" by Google one can't help oneself feeling set back a couple of decades – if not a couple of centuries.

> **Thesis 10: We need a Digital Enlightenment: New and original intellectual categories** that *will help us function as humans in a fundamentally changed world. This is vital if we are to reflect critically and productively about ourselves and our roles in a Digital Society.*

Mankind has stood at similar crossroads before. In the 18th century, for instance, ossified social structures and economic shifts forced intellectuals in Europe to fundamentally rethink the rules by which society functions. Caught between the classical logics of antiquity and medieval scholasticism on the one hand, ever-widening cracks in the old order on the other, free thinkers were forced to reconsider their basic beliefs and to adapt to change.

It was before this background that the German moral philosopher Immanuel Kant penned his famous tenant: "To think for oneself at all times – that is enlightenment". Today, the challenge remains the same.

Categories for a New Enlightenment

Through the process of enlightenment, societies in Europe and the New World were able to topple aristocratic regimes, release societies from the bondage of class conceit and guild rules, and formulate the basic Rights of Man that continue to apply today. By casting off their bonds, men and women in the Age of Enlightenment not only set in motion powerful economic forces, but in the process changed the way they saw the world – and themselves. Today, as digitalization and networking are again changing our perceptions, we once more need to summon up the courage and ability to think for ourselves in order to achieve our full potential and create a world in which we can live peaceful and productive lives. Humans, in fact, are very good at adapting to new and changing circumstances; that is why we stand as a species where we are today.

What we don't need is a protracted and rear-facing discussion, either with cultural pessimists or techno-romantics. Both can't help us reap the benefits from digitalization and networking. Neither pseudo-religious debates nor vague conjunctures, much less blissful optimism can help us here. Instead, let's talk about the real meaning of digitalization and networking, about their societal, cultural and economic effects and how we can invent the necessary categories and parameters to understand and deal with what is happening today. Neither the blind worship of rationality for its own sake nor uncritical devotion to the so-called laws of political economy and the psychology of the individual are called for. And tools such as mathematics, informatics, media and communication theory only take us so far. We are deep into uncharted waters, and we need new maps and compasses to guide us. With them, we will hopefully be able to collectively survey the world anew.

Chapter 2

The History and Future of Networking

The History and Future of Networking

The history of networking is more closely related to the history of ideas than it is to the development of technology. The inventor of the Word Wide Web, Sir Tim Berners-Lee, was acutely aware of this. His choice of the word "web" is certainly a deliberate attempt to liken the electronic networked he dreamed of to the intricate, interwoven threadwork constructed by arachnids. In fact, he went even further and deliberately included the many potential implications of his brainchild when he wrote in his book *Weaving the Web*:

> „The Web brings the workings of society closer to the workings of our minds. "

The similarity between social and neural networks is indeed striking, and we are surrounded by numerous proofs. Take industrial globalization, for instance, which could never have happened without the development of worldwide communication networks. This took place at roughly the same time most especially of course the aptly named "World Wide Web".

The same goes for events such as the global financial crises of 2007 or the rise of "superfast" stock market trading which takes advantage of tiny latencies within computer networks. Another good example is politics. The Arab Spring uprisings would have been over before they started without the help of Facebook and Twitter.

In every one of these instances social structures in modern society are beginning to resemble the complex structures of the human brain. The Web, in the meantime, is acting as a catalyst, leading us to develop social organizations that closely mimic our thought processes.

The phenomenal growth of the Web itself is of course the best evidence of the connection between brain structure and social networks. Never before did a mass medium evolve to full maturity as a "medium for the masses" in less time. In fact it only took the Web five years to become a kind of universal conduit for other media which use it as both a delivery platform and a communications channel. We users don't need to have the slightest inkling how the platform actually works; we just use it intuitively because it works the same way our though processes do.

Things become slightly more complex if we ask the question: How did Berners-Lee himself believe human thought processes work? Like many members of the Internet's "founder generation", Berners-Lee was the product of the psychedelic counter-culture movement that sprang up in the 60ies and 70ies in places like California, but also in Oxford where he was studying at the time. Cultural change was in the air, and popular icons combined with romance ideals, and Art Deco, but especially with the drug culture which had originally sprung up in Europe and was only later transferred to the United States. But the atmosphere Berners-Lee grew up in also had strong roots in Indian mysticism, or at least a watered-down understanding of Indian spirituality. We can therefore safely assume that Tim Berners-Lee, along with most of the true Internet pioneers, was strongly influenced by this mixed cultural environment.

The "zeitgeist" that shaped Berners-Lee and his counterparts was a wildly exciting blend of psychedelic rock from artists like Jimi Hendrix and Scot Joplin, elements of Indian yoga and meditation techniques, a dash of math and science and leavened with a perhaps naïve yearning for a better world (or at least for a better society).

Looking closely one still discovers traces of these cultural forces at work today in the Internet, as well as in the many different "network cultures" that evolved from it. At the same time, this shows how essentially primitive our understanding still is of the brain and self-awareness despite all the recent progress in cognitive sciences.

Networking without networks

There are many fables and sagas, notions and ideas from various cultures and traditions that testify to mankind's ongoing fascination with the concept of networking. Early instances include the Teotihuacan "Spider Woman" in pre-Columbian Mexico as well as the "Dreaming Paths" of Australian Aboriginals or the philosophical, ethical, and religious tradition of Chinese Taoism. In the 60ies, the fathers of personal computing and the Internet drew heavily on two traditional sources. As scientist and mathematicians they were influenced by European schools of thought and especially by Greek logic and mathematics. And as members of the counter culture movement many of them shared an interest in Indian traditions, on yoga and the philosophy of the Vedanta, one of the orthodox schools of Hindu philosophy (the term "veda" means "knowledge" and "anta" means "end"). Both were important for the development of the personal computer as well as the early Internet.

Many of the young Internet dreamers were drawn to the ancient Indian philosophy of the early texts known as the *Upanishads*; not only because of the obvious erotic and sexual connotations. "Tantra", of course, means "tissue" or "web" and can be loosely translated as "coherence" or "context". Mankind and the Cosmos itself are seen as part of one coherent fabric that connects the "chakras", which Tantrists believe to be energy points or nodes in the body, the psycho-spiritual constituents of living beings, through which the life force or "prana" flows.

Prana is believed to be responsible for the body's life, heat and maintenance. Tantrists think that this network is invisible to "normal" consciousness and can only be revealed to purified and enlightened individuals through the process of meditation. In the course of this experience the individual achieves the ability to quite literally "see the world with other eyes", namely, as related in one of the ancient Tantric texts, the Tripura Rahasya:

> *"Only a change of perspective might eliminate (the) error, as by habit everyone sees the world the way one has been conditioned to."*

This quote, not surprisingly, is attributed to the mythical original "guru of gurus", Dattatreya. "Guru" in Sanskrit means "teacher", but it is actually much more complicated than that. The syllable "gu" (गु) stands for darkness, "ru" (रु) for light. The guru therefore is one who brings light into the darkness of our everyday ignorance – in other words the "enlightener", a term that ties in neatly with the modern European concept of rational enlightenment!

In order to avoid falling into this fundamental error, the Dattatreya warns, we must change our own mental image of the world by reminding ourselves that the first question to ask about the act of thinking, even before describing the content of the thought, is this: Who is thinking? In other words: Who is the subject, the ego behind the thought.

The oldest Indian holy texts, the *Veda*, talk of the "Atman" as the true self and the "Brahman" as the universal, transpersonal being. They are the underpinning of the Advaita Vedanta school of philosophy founded by the great Indian thinker and reformer Shankara Charya.

According to this school, the Atman is the individual's essence. It is being constantly filled (from "within" as well as from "without") with (self-) knowledge, thus creating the ego. The modern German cognitive scientist and philosopher Thomas Metzinger calls this the "self-model of subjectivity" (SMT), an interdisciplinary approach to understanding and explaining the phenomenology of consciousness and the self.

According to the Advaita Vedanta, the Atman is the universal life-principle that is separated from the Brahman by "nothingness" (i.e. the thought process), while the ego and the material world are separated by the "Maya" or illusion (sometimes referred to as the "lack of knowledge") that is typical for the world as it is. The goal of mankind is to restore the original connections between these pairs of opposites, thus (re)creating an interconnected world in which Atman and Brahman become one and the individual obtains liberation, or "Moksha", through unity with the "One Supreme Self" – in effect becoming part of the ultimate network.

Echoing the ontology of the Upanishad philosophers and their appealingly formal logic set down somewhere around 1000 to 500 BC, the Greek philosopher Pythagoras of Samos defined the "self" as a ratio between numbers. It seems almost too much of a coincidence that the great Greek thinker and mathematician would base his "Mystery School" on the logic of numbers around the time the Upanishads were writing down their own theories for the first time. Previously they had only been transmitted as part of a long oral tradition.

The *tetractys* of Pythagoras is a mystical triangular figure consisting of four rows of numbers that add up to ten – the perfect number. The number one (also called the "Monad") is indivisible and therefore represents the first being, or the totality of all beings which is also indivisible. The number two (the "Dyad") represents the principle of "twoness" or "otherness", an essential precondition for the creation (or better: incarnation) of man and the universe. Three, the "Triad", is the noblest of all numbers as it is the only number as its sum together with those below equals the product both of them and of itself, thus representing awareness of the world around us which in turn is represented by the number four, the "Tetrad" – the number associated with the four seasons, the four elements, the planetary motions and thus with the Cosmos itself. Together, these mystical numbers lead from singularity to a plurality characterized by numeric ratios – a recognition that led the German poet-philosopher Heinrich von Kleist, two millennia later, to ask in his play *Amphitrion*: "Me? Which me?"

Like the Upanishad philosophers and Pythagoras of Samos, we too today have come up against a new phase in the development of mankind, a new epoch that calls for a new way of thinking, this time driven by globalization and communication. The Internet has already changed the way many of us live, learn, work and play, as well as how we communicate with others, and this is just the beginning!

As if that isn't enough, the "digitalization of the world" (or at least of our image of and reflection on the world) is also shifting, turning us as individuals into "others". In the process, we have been moved from the center of our individual universes to the periphery of a global network, thus forcing us to change our perspective, as foretold by the Tripura Rahasya.

Since the rise of mass media and mass production, humans can no longer be perceived to exist as complete individuals but instead form parts of a whole. "We are many", as the Viennese artist Egon Schiele once said; not just the many out there in the world and in our virtual networks, but multiples within ourselves, constantly being filled from within and without with the knowledge of the world.

What once seemed like different aspects on one personality (at least we could image ourselves that way) today form in the eyes and ears of the digital "others" simply different aspects of independent digital identities. These can take on an autonomous existence in the Net or at least appear relatively independent from their originators, and this despite the fact that they are dependent on (and determined by) a hitherto unheard of degree of interconnection.

It's interesting to speculate whether the emotions unleashed in present debates about infringed copyrights and stolen intellectual property in the digital age are due to bruised egos on the part of authors who are confused and angered by this disconnect between writers and readers. Be that as it may, the growing technical interconnectedness and interdependence just serves to demonstrate more clearly that we have always been, in our own eyes, the "many". As we become part of the network, the more we replicate ourselves. And in the process we discover new, additional "networks"; an "inner" neural network, and an "external network on the level of atoms and molecules, as well as technical networks, just to name a few.

This is hardly a coincidence. The success and dynamic growth of the Internet can be traced to the fact that we are finally realizing that our inner selves correspond directly to the world at large, a fact that we often intuitively felt but never could express or truly experience. And that is actually a rather comforting notion.

But what has really happened? And more to the point: What has happened to us on this round trip from the early development of human thought processes to their externalization, the „Technicum" described by former WIRED editor Kevin Kelley in his book "What Technology Wants"[4], and back again? A kind of mental three-step through the history of media may serve to illustrate this, starting from Gutenberg's "invention" of the printing press to the "electrification" of Media (and other communications technologies) we finally to their digitalization.

In their pre-technical form, media simply acted as the intermediary between otherwise unconnected worlds and unreachable spheres. Marshall McLuhan may have coined the phrase "magical channel", but

---

[4] See: Kelly, Kevin: What Technology Wants, New York (Penguin) 2010

that is what media always were; a way to connect with the gods, the ancestors and the natural spirits. This opened up a semantic space in which shamans and spiritualists, hypnotists and psychedelics could ply their trade, even after Freud and his followers began to challenge them with psychoanalytical explanations.

Of the many spheres once occupied by media, only the technical remain today, and in fact with the coming of "mass" media nothing else is conceivable. Or, as McLuhan argues, mass media, in order to be understood by the masses, must be abstractions of spheres they deal with every day and therefore robbed of their magic.

Until, that is, the Internet came along and together with other digital media gave new life to the idea of intermediation within the context of a global network. This is leading incidentally to a democratization of the role of the intermediary which once was reserved for specialists and "professionals" and now has become a shared experience in the new perceived reality of "many to many" communication. We must thank Tim Berners-Lee for pointing this out to us by showing that the Web is, in an abstract sense, simply an image of the human thought process.

Metcalf's Legacy

Many years before the Age of the Internet, a young American scientist and entrepreneur named James Metcalf had an idea that is still a major influence today. He understood that, while a single computer might represent a valuable tool, its potential usefulness would grow by leaps and bounds if it were possible to hook many of these "calculating slaves" together to form networks.

He himself had created an easy and extremely cost-efficient way to accomplish this through his invention of "Ethernet" architecture which allows computers to be connected to each other via a simple telephone wire and later through coaxial cables which substantially increased its throughput. While thinking about a way to turn this essentially modest innovation into a profitable business model he had the idea to cast it in terms of the economic benefit of the new technology. He therefore formulated what has since become known as "Metcalf's Law", namely:

"The value of a telecommunications network is proportional to the square of the number of connected users of the system."

While this may seem at first glance to be pretty abstract and technical becomes compellingly clear if cast in a simple example: If a dairy that sells ten bottles of milk to ten customers can, by some ingenious stroke of marketing and salesmanship, gain a single additional customer, they will henceforth sell eleven bottles of milk.

If a technology company (say a telecom company or an Internet provider) that has been serving ten customers who each use the service for one hour adds a new customer, then it will not simply be selling eleven hours of service, but much more. Why did the customer decide on this service provider in the first place? Because he or she wants to communicate with the provider's other customers, and the more they are, the more useful the service becomes. By adding customers, the provider can cause both old and new customer to use the system more frequently and for longer periods of time. Unlike the dairy, which can only grow linearly, a network can grow exponentially.

This "network effect" only works if each (new) actually causes the network as a whole to increase its utility. At the same time, each user profits from the effect. If not, the value of the network would decline. A telephone isn't "worth" anything in itself unless someone with whom I want to communicate suddenly wants one too. As more and more people with whom I want to communicate get a telephone installed, the greater the value of the telephone for everyone, both in the direct sense of economic value (dollars and cents) as well as in an abstract sense as communicative or personal value.

Since exponential growth is typical for the behavior of networked systems, Metcalf's effect does not obtain only to technical systems but to all forms. Take Google or Facebook as examples; each have experienced exponential growth over the past years, and the reason is clear: The perceived value of these networks for its users increases in relation to the number of "friends" I gain on or the number of relevant search results that are available. And of course, as Google and Facebook grow, their own relevance for advertising customers also increased exponentially.

Metcalf's Law has changed the way we understand and use networks in general, and nowhere is this more apparent than in the emergence of so-called "digital ecosystems". The term indicates what is meant: a system that bears more resemblance to a naturally growing "organism" than with a man-made machine. In a digital ecosystem, all the participants – vendors, customers, users, developers – are thrown together in dynamically shifting relationships and with fluctuating functions, similar to the way flora and fauna in nature engage and influence each other. In such a system, consumers are not simply paying customers but creators of valuable content and information. Users who make recommendations through so-called "affiliate networks" where they are paid commissions become an integrated part of the business model. And developers don't simply work to order; they create their own applications and thereby become full-fledged business partners, selling through "App Stores" or via Facebook.

Digital ecosystems are not only hothouses of economic growth; they also create new value and wealth potential. No wonder that managers the world around dream of creating such systems of their own, emulating such archetypes as Apple's digital ecosystem centered on the iPhone, the iPad, iTunes and its proprietary AppStore, as well as similar systems from Google and Facebook.

Life in the swarm

Unfortunately, these dreamers overlook a small but vitally important detail, namely that people who become part of such a "social" or peer-2-peer ecosystem are themselves changed by the experience. You only have to look at such successful examples as developer communities or any number of Open Source projects to see that individual users, while still remaining individually addressable, are in fact ceasing to act as individuals. Instead, digital ecosystems show many characteristics of animal swarms as they occur in nature.

Swarms of ants, birds or fish behave in remarkable ways. Though each member of the swarm may possess a very limited degree of intelligence, collectively they seem to act in highly intelligent ways. Also, they can band together quickly in order to perform a complex task and disburse just as fast. Ants for instance are very good at collecting food from widely separated location. In the morning, "scouts" move out and search the surrounding areas. Each ant leaves behind a trail of so-called "pheromones", a kind of scent marker. When the first ant finds a cache of food, it returns following its own pheromone trail. All the other ants remaining in the nest then follow this first trail and clean out the cache. By the time they return, the second scout ant will probably have brought back some food, and the rest follow its track in turn. This is a highly efficient method, and one that would require a computer to perform millions of calculations, in effect computing each and every possible combination, just like a chess computer is so good because it can calculate all possible future moves well ahead of any human competitor. Computer scientists are very excited about what they term "ant intelligent" systems, and lots of research is going on as we speak in the field of "swarm intelligence".

Humans, too, can act in "swarm intelligent" ways, be it teenagers on Saturday night using mobile phones and Instant Messaging to locate the best place to party or customers trolling through recommendation portals and social web sites in search of the perfect deal. Here, networking becomes separated from its technical rules and becomes essentially a lifestyle choice.

Swarms communicate, and they lack central leadership. Geese flying south for the winter are led by a single animal at the head of their v-shaped formation. That way, the lead goose provides uplift for those following, and the honk encouragement to their leader so he will fly faster, which slows them to conserve energy. When the lead goose tires it falls back into the rank and another takes over. Management theorists are very interested in this phenomenon, which they have already converted into the organizational concept of "vertical career paths": Instead of rising to the level of their own incompetence, as famously stated in the "Peter Principle", managers can move sideways into new jobs for which they in fact may not be fully qualified, forcing them to acquire new competencies and skills which in turn make them better leaders. Under the old hierarchical model, they would have probably been forced out because their upward paths happened to be blocked.

Mankind, it seems, has lots to learn from swarms, and the benefits for the individual can be tremendous. This doesn't stop critics like Jaron Lanier from abhorring this collectivist side of digitalization and networking. In an essay for edge.org he describes it as "Digital Maoism": "What we are witnessing today is the alarming rise of the fallacy of the infallible collective", he maintains. In short: While a swarm of connected human minds might be a fantastic resource for tracking down software bugs or discovering obscure gems on the Web, if you want to come up with a good idea, or a sophisticated argument, or a work of art, you're still better off going solo.

We have no beef with that. Of course individuals will continue to play important roles in art and argument. But we feel that Lanier's instinctive response to the collectivist aspect to networks is unreflected and above all typically American. (Full disclosure: one of the authors of this book, itself an example of a collectivist effort, hails from the United States.) The blind worship of the individual which is bred into the DNA of many Americans is no help in truly understanding the changes that digitalization and networking are bringing about and that appear to be both inevitable and highly desirable.

We believe that, as a swarm, customers are superior to enterprises that continue to behave both individually and egoistically. Collectively, customers display greater speed and flexibility, as well as intelligence and creativity; in short, the customer swarm possess more "vital energy" that the companies they deal with.

Enterprises will do well to tap into this source of vital energy by connecting closely with the customer swarm and in essence listening hard. This requires rethinking the role of such departments as marketing or advertising, which will have to be transformed into "inbound channels", effectively turning around the flow of information they generate: Instead of broadcasting out preformulated messages (ads, TV spots, etc.) to the widest possible audience, marketers and corporate communications professionals will engage in bringing back findings from the swarm so that other departments such as R&D, design or manufacturing can use them to produce new offerings tailored to fit individual customer needs. We believe that this will lead to enterprises entering into symbiotic relationships with social networks and the customer swarm which will reach far beyond anything we have seen until now; in effect creating a common digital ecosystem in which the enterprise simply executes decisions reached collectively.

Far from being purely technical phenomena, as was once believed, network effect are already influencing the way we think, as well as purchasing habits and social mores. The American economist Yochai Benkler describes this extremely well in his book „The Wealth of Networks"[5], in which he coined the phrase „Networked Information Economy" to describe organizations and individuals that do not follow the traditional rules of the market economy in which monetary profit trumps everything. In fact,

---

[5] Benkler, Yochai: The Wealth of Networks, YaleUniversity Press(2006)

large Open Source projects like the Linux operating system or the Firefox web browser are increasingly becoming the biggest competitors of such "market leaders" as Microsoft or Apple. The reason is clear: a "swarms" these project groups are superior to the large, unwieldy centralized structures these corporate giants rely on. According to Benkler, these collectivist organizations, by refusing to concentrate on the "bottom line", can offer the members of their teams "non-monetary rewards" that participation can bring. While these rewards will not always replace a paycheck, what he terms the "social-psychological component" of the reward already supports monetary appropriation for those engaged in creating the world greatest encyclopedia, Wikipedia, as well as the world's greatest tell-all website, WikiLeaks.

There are too many changes being brought about by network effects to discuss them all here. May it suffice to say that the genie is out of the bottle, and Pandora's Box is open. As a society and as individually, this means we will have to learn to put our inherent and inherited ability to think in a networked way in order to meet the challenges facing us in a world in which network effects and the information economy are rapidly changing almost everything.

The best way will be to learn from and emulate swarm intelligence. One important lesson is this: It makes no sense to try and stem the flood of information that in threatening to swamp us. Instead, we should go with the flow and acquire the ability to swim in like a fish, using its strength and vitality to assist us in reaching new shores.

In short: We need to find our way back to integrated ways of thinking about the personal and social reality of the 21st century. Fear will be as unhelpful here as trying to adapt old and tired concepts, rules and terms of reference forged in the Analog Age to fit the requirements of the Digital Millennia. At this point, it will be best to hark back again to Immanuel Kant, one of the fathers of the original Age of Enlightenment, and his famous dictum: "Sapere aude!" Think for yourself, he admonishes us, have the courage to think things anew and see them in a different light. And the wisdom to make the right choices as we head for the digital future. We need Digital Enlightenment today more than ever!

Chapter 3

Thinking in Real Time

Thinking in Real Time

Digital Enlightenment presupposes a new way of thinking that supplies us with the new terms of reference and categories necessary to understand what is happening. New schools of thought typically emerge when social, economic or other material factors change enough to make old terms of reference and old thought structures so out of date they can no longer help us make sense of the world around us. In other words: when the disconnect between what we see and what we think of as "real" gets big enough to change our living conditions, then we need to rethink our own life plans and adjust them accordingly.

In a previous chapter we mentioned Peter Sloterdijk's essay which is in fact a tribute to Rilke's famous poem with the cryptic title "Archaic Toros of Apollo", that ends with this straightforward admonishment:

"You must change your life!"

If not, life will change you, one is tempted to add with an undertone of warning – and not necessarily to your good. This is twice true for life in and with the Internet. This in turn means we need a new way of thinking, but even more it presupposes a new way of acting. Both are two sides of the same coin: We need to think and act more intelligently, but not only that; we also need to think and act faster, preferably in real time.

"Real time" is an expression used by engineers and computer scientists to describe an action that is simulated at a rate that matches that of the real process. Or as the Webster Dictionary defines it, "the actual time during which something takes place". The Internet can give us the ability to think simultaneously and synchronously, or at least it creates the necessary conditions.

At this point the difference between "classical" enlightenment and what we will call "Digital Enlightenment" becomes clear. Unlike classical enlightenment, its digital cousin can no longer rely on leisurely "ex-post" reflection about events – we simply don't have the time! The world around us is subject to digital acceleration, a phenomenon discussed in the introduction to this book. This means that in order to simply keep up with events as they happen we need to speed up the way we analyze what is going on around us.

More and more events in the real world call for direct and simultaneous responses. This can mean street protests in Istanbul that erupt seemingly overnight, or the NSA's program for collecting phone records and other personal data, aided incidentally by the very telecoms and IT companies we entrust with our data. If you think about it, events like these have always called for direct, real time solutions, only we lacked the means and the tools. Today we don't have to wait for some bulky petition to wind its way through the legislative process. Theoretically at least we are already capable of reaching democratic decisions in real time, thanks to modern software and the principles of direct democracy already enshrined in many Western democratic constitutions.

Which begs the question, even if we wanted or were allowed to, are we as humans mentally and intellectually capable of reaching decisions quick enough? Not too long ago the German chancellor Angela Merkel described the Internet as the "new frontier" – and that 20 years after the start of the World Wide Web! This is just about as far away from "real time thinking" as it is possible to imagine! Obviously, we still have lots to learn.

Digital Enlightenment, we believe, depends on our developing – or perfecting – our ability to evaluate in real time. Perhaps we should follow the Webster Dictionary's second definition of the term here, namely "the time it takes for a process under computer control to occur". Computers don't do one thing after the other; they "multitask", which is to say that they perform two or more tasks in extremely quick succession. This can actually slow down the time the computer takes to finish a certain task, but it increases its overall productivity. Which task is performed first depends on the priority assigned to it by the programmer. Getting our priorities straight in the Digital Age will presumably be just as hard as learning a new way of thinking.

Critics of digitalization like Frank Schirrmacher believe that the human brain isn't built for multitasking and that we therefore should be very worried about information overload. They are wrong. The thinking "subject" will always construct its own "self-model"; one that revolves around subjective experience. This model can and will be constantly adjusted as we process our perception of reality. It is reassuring to see that we seem able to change our way of thinking to better cope psychologically with the tectonic shifts in perceived reality we are experiencing in the digital world and through digital media. This is backed by decades of research in the field of cognitive science. Thomas Metzinger[6], a Fellow at the Gutenberg Research College in Mainz where he holds a chair in Cognition Enhancement, describes this in his book "Subject and Self-model" as follows:

"The direction in which evolution of mankind is moving has been reversed, if you will. In other words: If self-similarity is the goal, namely the ability to constantly improve our model of ourselves, then the result is a representationally optimized inwardness."

The challenge, therefore, is to adapt our image of ourselves to our fast evolving perceived reality – clearly anything but an easy task. However, the digital world itself offers us powerful tools that can help us solve the many urgent problems the networked world creates.

Take for instance copying, a simple, straightforward technology which proves that it is not always necessary to reinvent the wheel. Often, a "quick and dirty" solution that worked in a completely different con-

---

[6] Metzinger, Thomas: Being No One: The Self-Model Theory of Subjectivity, Bradford

text is enough. Nowhere has the "art of copying", so to speak, been more successful than in the digitally connected world. In fact, it's an essential part of this world, as Dirk von Gehlen[7] pinpoints in his book "Mashup – in Praise of Copying":

"The digital copy as a mode of reproduction blurs the borderline between model and imitation until original and copy are indistinguishable."

While it is no doubt interesting to dwell on the possibilities modern copying technology can offer (just imagine modern 3D copiers that can create real three-dimensional objects from 3D scans), it is even more fascinating to observe the changes that digital copying has wrought in our way of thinking, in our sense of personal morality. It used to be that a copy was considered inferior, if not downright disgraceful or even illegal. This was especially true in the realm of copyright and intellectual property; both concepts handed down to us from another age and both rapidly becoming obsolete. A copy, and especially one produced "quick and dirty", is usually not only the simplest but also the fastest solution: that is a value in itself in a "real-time environment". Just think of the advantage of having an edge on the competition when it comes to developing new methods of production, not to mention the cost advantage of 3D printing. Widespread use of these new technologies will presumably level the playing field for industries in the third world.

---

[7] Dirk von Gehlen, Mashup: Lob der Kopie, edition suhrkamp

Rilke got it right, it seems, when he poetically prophesized that it may not be enough just to change one's way thinking about the new reality; in fact we really need to change our lives. Both the adjustment of reality and of our own reflection thereon are happening as we speak, powered by the twin pressures of digitalization and networking. Both enable us to envision a new way of thinking, a new way of living and in the end a new way of perceiving reality.

But before we take off, let's first examine the new way of thinking required from us if we chose not to leave the development of the digital and networked future to others.

Why Digital Natives are not a new generation

When the protagonists of the new digital millennium talk about "digital natives", they aren't really referring to a "generation" at all; what they mean is a group of people who have already made progress in refining the way they think and their ability to understand and reflect upon the new digital reality around them.

Sociologists are fond of dividing human development and history into so-called generations which is slightly misleading because it assumes that a group of individuals born before or after some arbitrarily assigned date either belong or do not belong to the category being described. "Digital Natives" are a prime example.

The ability of the individual to adapt to a changing communications environment has more to do with how we experience and react to changes around us. "Digital Natives" therefore are not just those born after, say, and 1989; instead they represent those who have grown accustomed to dealing with digitalization and acceleration and who know how to use them to their own advantage. There are many people around who were born long before than 1989 and who feel perfectly at ease with the digital world. And there are obviously many younger people who have problems adapting to the future.

We believe that is no such thing as a digital native, at least not if the usual sense of attempting to describe a complete generation. Digital competence has nothing to do with biological age but instead depends on the individual's ability to recognize and adapt to changing surroundings and developments. Anyone attempting to understand a computer by using terms of reference from the Middle Ages will fall into the fallacy of describing it as a giant calculating machine. Such a description completely misses the point. Only by employing new terms and principles such as "data" and "data processing" can we truly grasp and appreciate what is going on in technology today.

This is less of a voluntary act than the result of self-analysis and perception of change around us; rather a kind of meditative attitude. More and more of us are currently gaining new experiences that must not necessarily lead to some kind of mystical or spiritual insight (although the potential is there).

For most of us these experiences simply lead to an expansion of what we perceive as "real". This can include sequences of numbers that follow a certain inherent order – so called algorithms. Three-dimensional objects can be part of this perception, as can pictures and scenes that consist solely of binary numeric code: essentially everything that we have learned to consider part of a "virtual" reality, as opposed to the "real" world. Of course all we experience as part of a scientific simulation or a computer games are digital objects, either in the form of digitized images of elements from our "normal" physical surroundings or as digital constructs from fantasy worlds.

By connecting these digital worlds and allowing digital information to flow freely, unhampered by the constraints of time and space, much less of political or geographical boundaries (admittedly still a long way off) we are in the act of changing or way of thinking. What this boils down to is that our thought processes are increasingly being decoupled from sequence, from our own personal histories and the history of society. In this context information is no longer a "scarce commodity" --- a concept that forms the basis for entire industries such as publishers and news media are dependent --- and instead become available anytime, anywhere and in unlimited abundance. Today, digital data are increasingly seen as a normal part of our reality as well as our conception of reality. Instantly available, easily duplicated without the need for any kind of special skills or knowledge, reassembled and reprocessed, digital "content" is increasingly becoming part and parcel of our new view of the world.

For this reason a growing number of people whose way of thinking is influenced by digitalization cease to understand why they should pay for news or news-gathering. The first are ubiquitous, the latter the job of algorithms or of individuals with the ability and willingness to share. No longer do we need trained and designated experts to work their magic: Gathering and disseminating information has become a normal function of modern, interactive media usage. They no longer require special knowledge or abilities in a networked world. Instead, they have become a fundamental cultural skill, like reading, writing and arithmetic – something every child in modern society learn from the very beginning.

How digitalization changes our perceptions (of the world and of ourselves)

Knowledge in this world has less and less to do with specialist expertise, or, as *Guardian* author Mercedes Bunz notes in her beautiful book "The Silent Revolution"[8]: "knowledge tends to swarm all over". This contributes to the fundamental shift in our perception of reality, something that includes our own self-view, as modern cognitive science teaches us. How

---

[8] Bunz, Mercedes: The Silent Revolution: How Digitalization Transforms Knowledge, Work, Journalism and Politics Without Making Too Much Noise (Palgrave Macmillan) 2013

this comes about is described by Bunz in such as fascinating way that we have decided to paraphrase her essay here in order to explore some of her conclusions and their consequences.

In the "digital space", Bunz maintains, the nature of knowledge itself is changing:

- "Instead of one authority that affirms something as a fact, there is now a choir of voices, in whose plurality information needs to stay consistent to be considered a factual..."
- Vast databases offer a knowledge way beyond everything the memory of an expert could possibly know, and makes it more likely that experts might miss something.
- Digital information constantly produces news, and facts change fast. This makes expert knowledge supposedly outdated and not accurate."

This creates something that transcends what we normally describe as media: a new kind of "information sphere" that follows us around wherever we go, thus becoming an important part of our perception of reality. In the past, reality often ended up as media content; now it's the other way around.

Thanks to digitalization, we can share and process this medial reality digitally. What we now think of as "interactivity" enables us to do this effortlessly and quickly. As a result, we are coming to a new understanding of the validity of information. Or as Mercedes Bunz put it, "For the objective reporting of an event, journalism isn't the only guardian of our social truths anymore but find at its side the digital public with its choir of voices." As the witnessing of the digital public happens instantaneously (and thus without any chance of vetting), the course of events "needs to be extracted from the sameness found in the plurality of voices." The fact that a news item keeps cropping up from a variety of sources becomes the main reason we believe it's true."

For the digital public, specialists and witnesses are less important than active participants and other members of the community or of a social network. Her conclusion from all this echoes the title of this book. "Focusing on the social force of digitalization, it can be said that we are entering a second phase of enlightenment and emancipation: after the autonomous subject, whole crowds are asked to form themselves under the banner of enlightenment: *Sapere aude*" ('Dare to know!)."

Mercedes Bunz, too, calls for us the change of perspective, all the more because the reality surrounding us as we work and play itself has changed fundamentally. We have no choice but to switch our viewpoint if we don't want to lose our touch with this changing reality.

Probably the most radical proponent of this concept is the American activist and author Kevin Kelly, a former journalist, photographer, founder of the *Whole Earth Catalog* and later one of the publishers of *Wired*, the central organ of the digital avant-garde.

His book bears the slightly unsettling title *What Technology Want*.

But how, we ask ourselves, can technology "want" anything? Which in turn begs the question, is technology an autonomous living system that, like humans, possesses consciousness of its own surroundings and is capable of developing free will? Not even someone as radical as Kevin Kelly would go so far. He does; however, direct our attention to the often overlooked fact that biological systems such as humans, but also technology developed by humans share a crucial attribute: "Both seem to be based on immaterial flows of information."

For Kelly, technology, like art, is characterized to a large extent by the information they contain and less by their outward material form. Therefore it is possible for him to describe technology as an independent, autonomous "sphere" of reality that exists separately from its human designers. Art, music and other intellectual, immaterial creations also belong to this sphere, as they, too, are largely characterized by the information they contain. He coins the term "technicum" to describe the globalized, interconnected stage of technological development. He argues that the technium a power of nature and that the processes that create it are akin to those of biological evolution. Not that we have to do everything the technicum tells us to, but we should learn to harness its power to our own advantage instead of fighting it. "Seeing our world through technology's eyes has, for me, illumi-

---

[9] Kevin Kelley: What Technology Wants, New York (Penguin) 2010

nated its larger purpose", he writes." And recognizing what it wants has reduced much of my own conflict in deciding where to place myself in its embrace."

So for Kelly, too, it is important for us to be able to change our perspective which, as it turns out, is a basic human ability. And while Kelly strives for a general "technological" perspective, we believe that it is vital to adopt one that encompasses both digitalization and networking as part of our outlook on the world around us. The goal is not to find an "ultimo ratio" for the world which is changing much too fast for that; instead we suggest finding a method that will allow us to perceive reality in a different way, enabling us to make sense of it. The methods of the past are no help here. Ultimately we need to understand that we have to change ourselves, our ways of thinking as well as our perspectives in order to grasp what is happening in the digitally networked world. And we need to do it in "real time" while the new is still in the process of forming itself – and before others take over control.

Why "multitasking trauma" is just a myth

In Kevin Kelly's scenario, the "technicum" takes over the role of awareness as a kind of a technical perception, a new dimension of existence that is brought about by technology and constantly being refreshed and recreated. This new dimension transcends our material situation but also reflects it, thus developing it further.

Digitalization and networking take this process a step further. "Virtuality", a sphere between (material) reality and (non-material) perception of reality that adds a new dimension to what we think of a "real", has moved from a fixed idea in the heads of a small group to a commonplace phenomenon felt and observed by many. Am I less "real" for my circle of friends on Facebook than for acquaintances on this side of the computer screen? Hardly: We need to learn to think about this additional dimension in new ways if we want to be able to shape our future. Neither Google nor the NSA will stand still and wait for us to catch up; the faster we learn to change our lives and our perceptions of reality the better.

However, in order to understand this changed world we need new categories to describe and think about them. Without proper terms of reference (without categories, in other words) we won't be able to truly understand what is happening. According to the principle of linguistic relativity first proposed by 19th-century thinkers such as Wilhelm von Humboldt, language determines thought, so linguistic categories limit and determine cognitive categories. Quantum physics, as Niels Bohr famously put it, is not concerned with reality per se but with "what can be said about the world".

The American theoretical physicist John Archibald Wheeler (1911-2008) suggested that information is fundamental to the physics of the universe. According to his "'it from bit" doctrine, the laws of physics can be cast in terms of information, and all things physical are information-theoretic in origin. This in turn caused Christopher Langan, a self-taught philosopher and physicist, who has been called "the smartest man

Digital Enlightenment

in American because of his alleged IQ of 195, to describe the universe itself as both "conscious" and "introspective", a state he calls "infocognitive".

In order to really understand the world, we therefore need to include and determine its informational content. Information ("Bits" in Wheeler's dictum) is criteria for proving it something is "real".

Which proves that multitasking, namely the ability to handle more than one task at the same time is not, as Frank Schirrmacher would have us believe in his book "Payback", a traumatic form of bodily injury but a way of handling digital reality. Schirrmacher himself, while by his own admission a regular user of digital media and networks, apparently doesn't belong to those who are willing and able to develop this new skill that would have enabled him (he passed away quite suddenly in June of 2014 at the age of only 55) to help understand and shape the digital future and think in new ways. In this he shares the fate of Jaron Lanier, an American musician and self-proclaimed "father of virtual reality", who persist in seeing in the swarming of information and the creation of the new "digital public" something that he calls "digital Maoism".

Our brains have always been quite capable of performing multiple tasks more or less simultaneously. The simplest example of human multitasking is of course the ability of most of us to take a phone call while typing an e-mail. Our feeling of "hunger" persists in the background until we eat something; we cannot turn it off at will. However, the ability to multitask seems to be more marked in members of the younger generation who have grown up with digitalization and networking as a determining factor of their lives. Multitasking may even prove to be a crucial evolutionary development by our species which can stop homo sapiens from being marginalized by its own technology.

As the cognitive scientist Linda Stone has noted, kids who grow up in the digital world are especially adept at multitasking, but pretty lousy at concentrating fully on a single task (hence their tendency to keep the TV set on in the background while they do their homework, at the same time keeping an eye on their e-mail or WhatsApp inbox).

Schirrmacher describes multitasking as a form of Taylorism, namely the breaking up of complex tasks into many tiny, dull sequential steps. In this he echoes Aldous Huxley, who in his dark future scenario "Brave New World", written in 1932, coins the term "Fordism" to describe a world in which everything is organized along the principles of strict division of labor and assembly line production.

Digitalization and networking point us in a totally different direction, as will be seen later. Instead of a world of mass production the result will be a world of digital craftsmanship and (networked) manufacturing – a term which comes from the Medieval Latin *manufactura,* and means literally "made by hand". Developing the ability to keep up with the clock pulse set by our networked computers, so in effect to lead our lives in "real time", we feel will be crucial if we hope to be able to stay in control.

The ability to do so is inherent in all of us; it is in fact hardwired into our brains. The task we now face is a social one, namely to bring as many of us "up to speed", so to speak. School children need to be instructed in the use of digital tools and made able to perform tasks in real time. Algorithms and computers, networks and applications that depend on them will help us, but it is up to us put them to creative and productive use. Thinking in real time is rapidly becoming a fundamental cultural skill, and by doing so we have a chance of controlling and mastering our futures. If not, we risk being overrun by developments so dynamic that we will no longer be able to stay the course.

Chapter 4

The Networked Human

The Networked Human

The Internet changes everything, so why not its users, too? It is this aspect of digital acceleration and change that cultural pessimists hate the most. Yet in fact it would be surprising if it were otherwise. All you need to do is look back on the evolutionary history of mankind to see that our ability to adapt to changes in our communications environment is what really makes us human.

The present generation has been born into a medial environment that is already marked by digital acceleration. Processes that once took time now can be performed in an instant. Kids are oblivious to this since they never knew anything else.

If Jaron Lanier, Nick Carr and their ilk were right then young people should be the first victims of "digital dumbing down", meaning that they should be exhibiting symptoms of linguistic and cultural degeneration. Their language skills should be measurably poorer than elder people's, their literary abilities inferior to those of their parents and grandparents. True: The mother of German thinker and author Ernst Jünger could recite Goethe's entire *Faust* by heart, an ability that is rare today. And we are told that audiences in the time of Aeschylus and Socrates often could recite entire plays after hearing them only once; such talented individuals were often enslaved and sold as private teachers to Sicily.

But let's not forget that in the early days of the Internet in the not-so-far-back 70ies and 80ies cultural pessimists, most of them teachers and philologists, warned against the degenerative effects of electronic media such as radio and television. Writing skills among young people would decay, they believed. Radio and especially television, the proverbial "boob tube", were going to produce a disenfranchised generation (ours!) who would become helpless captives, mindless zombies, capable only of reaching for their remote controls to surf through mind-numbing soap operas, crude actions film and – this was still very new and hence even more perditious – reality TV shows.

The result, these critics argued, would be a rapid decline in literacy. "Kids no longer write letters", language instructors would murmur in antediluvian tones.

It is to presume that these same critics are absolutely delighted to hear that their gloomy predictions have failed. In fact by 2012, according to an estimate by the Radicali Group, more than 145 billion e-mails were being sent every year, which means that, given a world population of some seven-odd billion, every one of us including children and the aged write more than 20 e-mails a year on average. This number is expected to grow to 200 billion by 2016. So much for the end of literacy!

In fact, given that large parts of the world population still lack Internet connections, it seems that young people have never written as much as today. Yes, they don't necessarily write in the style most English professors would prefer. They are, however, at least as articulate as the last generation and possibly even more so since they have added to their vocabularies a host of acronyms and abbreviations such as "LOL" (for "laughing out loud") that are well understood and which add value to written communication, as do these curiously usurped groups of punctuation marks — "smilies" which consist of semicolons, commas or brackets which, when turned sideways, look a lot like smiling or frowning faces and which are understood by all in their peer group.

Game nuts are really quite peaceful

Another popular scare topic for digital pessimists as well as for the classical news media is the supposed brutalization of the young generation. We are falling back into barbarism, they say, and it's all the fault of shoot-em-up games, violent videos and of course the Internet. No wonder kids steal their dad's guns and go on a rampage, like James Eagan Holmes who shot 12 at a school in Aurora, Colorado, in 2014, or Anders Behring Breivik who murdered 69 teenagers aged 14-19 at the Norwegian island of Utoya in 2011.

Not that this form of mass paranoia is new. In our childhoods commix were the root of all evil, especially the really gory ones. Young people were certain to turn into mindless monsters, out parents and grandparents believed. But unfortunately for the preachers of doom there are no facts to back it up. Instead, adolescent violence in the 50ies and 60ies was in steep decline, as official statistics prove.

Today newspapers and TV channels are full of stories about mass killers and of youth gangs beating up innocent bystanders in subway stations while dozens of commuters stand by and watch. Clearly, at least Joe Public thinks so; the propensity for violence among young people is on the rise today, especially those with migration backgrounds. In fact the German Federal Police, in a press release in 2013, noted that "the falling number of adolescent suspects is significant. This is especially so in the 14 to 18 year-old age group where the number of suspects in cases involving violence crimes has fallen by almost 9 percent … while those involving grievous bodily harm have gone down by 9.4 percent."

In 2008, a pair of economists, Gordon Dahl and Stefano Della Vigna, published a paper in the findings of the National Bureau for Economic Research entitled *Does Movie Violence Increase Violent Crime?* in which they found that violent crime decreases on days with larger theater audiences for violent movies. They explain this explained by the self-selection of violent individuals into violent movie attendance, leading to a substitution away from more volatile activities. In other words, instead of prowling the streets looking for trouble, these kids are sitting in movie theaters eating popcorn and drinking soda while watching celluloid brutes beating and shooting each other.

Our above-mentioned friend Norbert Bolz, professor at the Technical University of Berlin (see chapter 1) believes that playing digital games is the "high road to the digital future". In his view, media competency is won but by reading manuals but by having fun with programs. He writes:

"Each sense of achievement marks a new border crossed in human-machine interaction. Only those who know how to play with a computer can hope to use them as tools – and have fun at the same time."

Evolution in fast-forward

Humans have always lived in symbiosis with their tools. By inventing more and more sophisticated and intelligent tools, mankind has changed itself and become smarter over time. This is a natural process, and it is nonjudgmental. The mistake the prophets of doom make is to assign value to it. One typical example is the belief in "progress", as though mankind were striving throughout its history to reach a certain goal that will make them better humans. Another is the worry about technology-induced human degeneration, or as Frank Schirrmacher once put it in an interview, "my brain is being squashed by the Internet".

Ray Kurzweil is a visionary as well as a serial entrepreneur which gives him sufficient financial independence to pursue his investigations into the way technology is shaping human destiny. He is one of the founders of the so-called "singularity movement" which can be traced back to the early computer scientists John von Neumann in the 1950s and *Irving J. Good* in the 1960s and which hopes to achieve man's age-old dream of immortality. Singularity describes a world in which mankind, far from being dominated by machines, will become one with them, melding together and thus producing new, superior kind of intelligence which can help become better and more "human" beings.

Kurzweil first came in contact with singularity in the 80ies at about the same time he first heard of *Moore's Law*. Gordon Moore, one of the founders of Intel, predicated that the power of semiconductor chips would double every 16 to 18 months. Doubling down, as every gambler knows, leads to enormous acceleration. We call this "exponential growth".

The oldest story about exponential growth comes from ancient India, where it is alleged that the inventor of the game of chess was told by his grateful sovereign that he could have a wish. He asked for as many grains of rice as would be needed to fill a chessboard, with one grain on square one, two on square two, four on square three, eight on square four, and so on, doubling the number of grains on each succeeding square.

The sultan immediately agreed to such a modest wish but found to his dismay that he had just given away 18,446,744,073,709,551,615 grains of rice – much more than the royal granary contained or that probably even existed in the world. Simple: by doubling 64 times, you reach the number 1.84×10^{19}. Whether the sultan proved true to his word or whether (which seems much more likely) he just had the inventor of chess quietly beheaded, history does not record.

Kurzweil and the adherents of singularity believe that somewhere around the year 2020 we will be able to build machines capable of emulating the human brain at a cost of less than a thousand dollars per computer. That would eventually allow us to create a "backup" of our memories and experiences and thus, for all intents and purposes, making us immortal.

But singularity, he conjectures, will do much more: enabling us to create tiny artificial cells which can replace older cells as they wear out, effectively stopping the aging process and granting us eternal youth! In his book *The Age of Spiritual Machines* Kurzweil writes that it will be possible to scan brains from the inside using "nanobots" and then reverse engineer them in order to understand and emulate human thought and memory. This "non-biological intelligence", as he calls it, will combine with our natural intelligence to enhance our cognitive abilities and create "super-intelligence". He points to recent developments in the field of neural implants which he thinks will one day make Moore's Law applicable to human intelligence, doubling our IQs every 16 to 18 months...

If this sound like pure science fiction to you, then we suggest you visit Moffett Field in Mountainview, California. There, at NASA's *Ames Research Center*, the *Singularity University* has its campus in a huge former dirigible hangar.

Singularity U was founded by Google's Larry Page, among others, and is maintained through generous grants from a wide number of Silicon Valley entrepreneurs. It only takes 80 new students per year, and the waiting list is long. For $25,000 managers can sign up for so-called executive briefings which consist of lectures by some of the leading lights in the fields of nanotech, biotech, artificial intelligence and robotics.

The sixth Kondratieff

Participants hope to learn about the "Next Big Thing" in technology, the next stage of the rocket, so to speak, that will propel us and the economy to the next higher plane.

Managers and economists have always looked for inside information about the future. One of them was the Soviet economist Nicolai Kondratieff, who was put to death on Stalin's orders in 1938 and who is hailed as the inventor of the "Kondratieff Cycles". These describe economic growth in terms of *supercyles* of 30 or 40 years, each triggered by a "basic innovation" that launches a technological revolution that in turn creates a new leading industrial or commercial sector.

Adherents to this theory claim to recognize five such supercycles, starting in the 1770ies with the Industrial Revolution, followed in the 1830ies by the age of steam and railways, the Age of Steel and Heavy Engineering from 1875, the age of oil, electricity, the automobile and mass production from 1908 onwards, leading up to our current age of information and telecommunications which began in the 1970ies.

If Kondratieff was right, our current age is now coming to an end, and the cognoscenti want desperately to find out what the basic technology in the next supercycle will be. Many bets are on m biotechnology, ushering in decades of health and wellbeing.

Whatever this *Sixth Kondratieff* will be, Kurzweil believes it will be one in which singularity comes to the fore, a technological Fountain of Youth.

Of course, not everyone subscribes to this optimistic view of the future. Theodore Kaczyinksi, who became infamous as the "Unabomber" who held America breathless by sending a series of letter bombs to celebrities and government officials, killing three people before he was caught in 1995, called in his "manifesto" on mankind to revolt against the alleged "industrial-technological system" which he called "a disaster for the human race" because it would turn us all into the slaves of our machines. Which harks back to Mary Shelley and her 1818 sci-fi thriller "Frankenstein", the story of a young German scientist who create a human-like being who unfortunately turns into a monster?

And in more modern times, popular TV series include *Fringe*, in which a team of scientists uses fringe science and FBI investigative techniques to investigate a series of unexplained, often ghastly occurrences, which are related to mysteries surrounding a parallel universe. The series has been described as a hybrid of *The X-Files*, *Altered States*, and *The Twilight Zone*.

Making a map of the brain

Just how far we have come along the road to singularity is demonstrated by Dr Christopher deCarmes, the CEO of Omneuron, a company based in Menlo Park in the heart of Silicon Valley which specializes in developing machines that are capable of visualizing and controlling the functioning of the brain using non-invasive methods based on Magnetic Resonance Imaging. "Reading someone's brain" is probably how most of us would describe it. His machines are used to combat depression and to measure pain thresholds, and the point the way to the future of neurology. "We are already able to read information generated in human brains", he says.

This means that sometime soon we will probably be able to turn the process around and reprogram the brain. Optimists see in this a kind of Nuremberg Funnel (German: Nürnberger Trichter), a jocular description of a mechanical way of learning and teaching. Others worry that this technology will create legions of robot-like work slaves who are remote controlled by their digital masters, condemned to blindly follow the will of computers and their human programmers.

Kurzweil and his associates, understandingly, see the future in less dramatic terms. He speaks of "accelerated evolution and maintains that "technological evolution is an outgrowth of – and a continuation of –

biological evolution", in other words a completely natural process.

We may be closer to this then we think. In 2011, Craig Venter, the pioneer of human genome sequencing, first succeeded in creating life in the laboratory with his team. Experts believe that college students will one day be able to engineer artificial bacteria in biology class. And the U.S. government has pledged to spend at least 300 million dollars over the next ten years on the "Brain Activity Map Project" which is tasked with mapping the activity of every *neuron* in the human brain. It seems that scientists to this day do not fully understand completely how thought and memory are achieved. If we did, we would probably soon be able to develop artificial brain cells than can be implanted in humans.

At first, of course, these cells will be used to treat degenerative diseases of the brain such as Alzheimer's. But who is to stop us from injecting these cells into healthy patients, thus potentially producing "super intelligence". If so, this raises a number of tricky questions for society, such as, "Who gets the extra cells and will my health insurance pay for them?" Or will super intelligence be only for the rich? Will society become divided into a class of privileged geniuses and a majority of "average intelligent" dummies?

The long-overdue discussion about Digital Enlightenment which after all is what this book is about must necessarily include a very public discourse on rules and morality in an age of digital transformation and acceleration, because developments such as those just described naturally hold a high level of destructive potential. If we let it happen to us, then we truly deserve to be called "digital dummies"...

In praise of distraction

Nothing enrages cultural pessimists more than when you tell them that multitasking is actually good for you! They think that doing more than one thing at once is a sign of degeneracy and that that kids are getting dumber due to digital technology. Not so, as psychologists and researchers like Linda Stone tell us. She coined the phrase "Continuous Partial Attention Syndrome", or CPA, back in the 90iers, and she describes it as a necessary way of adapting to an increasingly networked world in which we humans function as nodes, simultaneously receiving and transmitting information within our sphere of family, friends, colleagues and acquaintances.

Kids in fact are becoming increasingly proficient at subdividing their attention in order to perform various tasks sequentially or near-simultaneously, such as doing their homework while listening to the radio and monitoring the e-mail, Twitter and smart messaging accounts. Compared to true multi-tasking, she believes, full attention is not required by CPA (hence the "partial") and the process is ongoing rather than episodic (hence the "continuous").

A human being capable of performing a number of task at once is, of course, at an advantage in a networked world where lots of things happen to us all the time: attempts to communicate with us, bits of information fired at us from all points of the compass, ways to entertain us ours for the asking. If you live in a network, than the ability to function like a node in that network, as Linda Stone describes it, is obviously an advantage – possibly even an evolutionary advantage, which would explain why it seems to be arising in its most pronounced form among the younger generation. "We want to connect and be connected", she writes "We want to effectively scan for opportunity and optimize for the best opportunities, activities, and contacts, in any given moment. To be busy, to be connected, is to be alive, to be recognized, and to matter."

Survival of the fittest, or at least the most adaptable, hinges more and more on what computer scientist call "network management", which essentially all about command and control. It is this talent that Stone and her colleagues think the next generation can already be seen to develop. Of course, there is a tradeoff: Kids appear to lack the ability to concentrate fully on a single task. This, of course, worries their parents, who take them to doctors who prescribe pills against ADHD (Attention Deficit Hyperactivity Disorder), a condition that Thomas Armstrong, Executive Director of the The American Institute for Learning and Human Development, calls a "myth". Even Dr Robert Spitzer, professor of psychiatry at Columbia University in New York and the psychiatrist who identified attention deficit disorder, admits that "up to 30 per cent" of youngsters classified as suffering from disruptive and hyperactive conditions could have been "misdiagnosed."

In a much-discussed opinion piece for the *New York Times Sunday Review* entitled "A Natural Fix for ADHD", Richard Alan Friedman, professor of Clinical Psychiatry at Weill Cornell Medical College, has a fascinating explanation for what seems to be going on among younger Internet users, and if true it means that CPA has always been a psychological trait of some, but not all of us. "Recent neuroscience research shows that people with ADHD are actually hard-wired for novelty-seeking", which he believes is "a trait that had, until relatively recently, a distinct evolutionary advantage." Compared with the rest of us, he says, they have sluggish and underfed brain reward circuits, so "much of everyday life feels routine and under stimulating."

Buried deep within our brains is a mechanism which causes the release of dopamine, a neurotransmitter, which plays a major role in reward-motivated behavior. In some of us, dopamine is produced by new and exciting experiences. Drugs, alcohol and sex can act this way, but also the constant pressure we experience from waves of information, e-mails and other content on the Internet that essentially dispel boredom and make life exciting for those whose otherwise feel under stimulated.

Friedman takes us back to the early days of homo sapiens: "Humans evolved over millions of years as nomadic hunter-gatherers. It was not until we invented agriculture, about 10,000 years ago, that we settled down and started living more sedentary — and boring — lives. As hunters, we had to adapt to an ever-changing environment where the dangers were as unpredictable as our next meal. In such a context, having a rapidly shifting but intense attention span and a taste for novelty would have proved highly advantageous in locating and securing rewards — like a mate and a nice chunk of mastodon. In short, having the profile of what we now call ADHD would have made you a Paleolithic success story."

Research appears to indicate that the number of people belonging to the "hunter" part of our species is growing; at least if we accept that ADHD is an indicator. Recent increases in ADHD rates have been explosive, as well as its treatment through medication. According to the Centers for Disease Control and Prevention the lifetime prevalence in children has increased to 11 percent in 2011 from 7.8 percent in 2003. 6.1 percent of young people were taking some ADHD medication in 2011, a 28 percent increase since 2007. Even toddlers are fed these pills to "quiet them down" and, presumably, to keep them from getting on elder people's nerves.

But if Linda Stone is right, we are actually hindering "hunter-types" among younger people from adapting to their new communications environment and thus placing themselves at an advantage over their duller, more sedentary peers. Instead of in terms of generational conflict it would be wiser to accept that there are two distinct groups of individuals in society --- hunters and gatherers, as Linda Stone would have it, Richard Friedman's dopamine junkies and dullards. By enforcing and rewarding the ability to concentrate and punishing (or doping) those who can't, society may actually be committing a serious crime on its young (or, if you prefer, a sin against itself) by interfering in natural selection!

Neuphobes and neophiles

Not that all members of the younger generation are hunters. In fact, instead of dividing mankind up into post or ante Internet and creating a potential generation gap, it is probably more helpful to follow Robert Anton Wilson, the American bestseller author, philosopher und anarchist, who chose the designation "neuphobes" and "neophiles" to explain why some of us are drawn to technology while others remain immune or even adverse to technical innovation.

The new, for many, is fraught with angst. Wilson, on the other hand, cites the famous admonishment of American pioneers, "go west young man", as a typical example of neophilia or the love of new things.

Wilson's ideas, which he first presented in his collection of essays entitled *The Illuminati Papers*[10], bear an uncanny resemblance to what is going on today in the realm of Digital Enlightenment. And Charles Darwin, the father of modern evolution, recognized in the 19th century that mankind isn't very good at anything except adapting to change. Evolutionary biologists ever since have been at pains to explain human superiority in terms out ability to adjust, thus maintaining a

---

[10] Wilson, Robert Anton, *The Illuminati Papers* (Ronin Publishing) 1980

genetic lead over other, perhaps stronger but less adaptable species.

Evolution, it appears, does not necessarily have to proceed in slow motion. At times it can fast-forward: biologist Andrew H. Knoll[11] talks of something he calls "permissive ecology" which can at times speed up natural selection. A good example is mass extinction, for instance the end of the dinosaurs due to a collision of a meteor with earth which created room and breathing space for a population of completely different animals (in this case mammals) who were better equipped to survive and spread out rapidly and evolving into new species in what was, at least for evolutionary biologists, just a blink of an eye.

All of this predates digitalization, of course, but it demonstrates convincingly – at least to everybody but confirmed cultural pessimists – that humans are able to adapt to new surroundings very quickly.

Not that we ourselves feel this. A team of Harvard psychologists, in a paper published 2013 in *Science* explored what they call the "end of history illusion," a widespread phenomenon which causes us to believe our personalities, after a certain age, are somehow set in stone. According to their research, which involved more than 19,000 people ages 18 to 68, the illusion persists from teenage years into retirement. When we remember our past selves, we usually see how much our personalities and tastes have changed.

---

[11] Andrew H. Knoll: *Life on a Young Planet* (Princeton University Press) 2004

But when we look ahead, somehow we expect ourselves to stay the same.

We don't. And even if the results of this research seem to confirm our almost boundless ability as humans to overestimate ourselves, there remains nothing else for us to do in the end than accept change as a constant in our lives. Today, in a world that is becoming more and more digitized and interconnected, we must recognize that there is room for everybody willing to adapt and grasp the opportunities that present themselves.

Divvying people up into post and ante Internet generations doesn't help at all. As always when evolution occurs there are those who are particularly adept at making the most of a new situation and others who lag behind. In evolution, there are always winners and losers, so why not when it comes to Digital Transformation?

The good news is that, in a networked world, individuals who are a step ahead get the opportunity to share their experiences with others and reflect them back to those having difficulties keeping up. Which does not mean that their won't be digital laggards who are unable to see how the world around them is changing or unwilling to understand that they, too, need to change if they do not want to be left behind in the analog past.

So how many of us are neophiles, how many neophobes? It's hard to find really reliable figures, but perhaps finding from adjacent fields of study can help us here. Motor Presse Stuttgart, a major German publishing house where one of the authors once worked, have long been accustomed to creating magazines for what they term "technically affine" audiences. Their market researchers long ago discovered that it is possible to separate potential magazine buyers into two groups who they call "TAPs" and "TRPs". TAP stands for "Technically Advanced Person", TRP for "Technically Retarded Person".

In focus group after focus group, the distribution of TAPs and TRPs turned out to be virtually identical: Anywhere between 40 and 50 percent auf adult readers were positively inclined towards technology, in this case cars, motorcycles, stereo equipment, cameras and photography or flying; all subject covered by magazines the company published. And while most of those survived identified themselves as "car buffs" or "audiophiles", as the case might be, all were at least superficially interested in all the other technical topics discusses in the groups and could, through gentle persuasion, be brought around to buying a magazine covering technical products they initially weren't very interested in.

This also indicates that about half of us are not "technically affine"; as they say in marketing speak. For us this means that the number of "digerati", that is those persons well versed in computer use and technology, are just about balanced in society as a whole by those who are either indifferent to or openly agnostic about technology. In terms of Digital Enlightenment, this means simply preaching to the choir will not be enough: We need to convince substantial numbers of non-techies to join us or at least help the stragglers keep up with the rest of us. This, we believe, will be a task for a new breed of social workers whose job will be similar to todays' therapists aiding and assisting those with different kinds of handicaps.

In any case, Digital Enlightenment has to do less with age than inclination. And since it's not about technology so much as about social and political change, it doesn't take a technician or a teenager to be able to play.

Everybody's gone surfin'

On the other hand there does seem to be some kind of correlation between the willingness of an individual to deal with things digital and one's age (or better: one's stage of development). An annual study commissioned by the two German public broadcasters, ARD and ZDF, showed that in 2014 eight out of ten Germans were more or less regular users of online media such as Internet or Web TV. In 2000, incidentally, this number stood at 26.6 percent. Men are more prone to surf the Web (83.7 percent) than women (74.6 percent). But these figures don't say much about the relative online presence of old and young. Taking a closer look, however, reveals that among the age group 14 to 19, there is virtually no one who is still offline; at least statistically 100 percent of all kids have been online users since 2010. Among those aged 20 to 29, the percentage is 99.4, among 30 to 39 year-olds it is 97.4 percent. The number falls slightly to 93.9 percent for those between 40 and 49 and more steeply, namely to 82.1, among the 50 to 50 age bracket. What really pulls the statistic down is final age group who are over 60, where only a dismal 45 percent admit that digital media and the Internet are important for their lives.

Of course, we grey panthers don't have long to live, and that will set the statistics right. Those of us born after, say 1955 will all get to see a world in which virtually everybody goes online as a matter of course. Give it ten more years or so. By then, use of digital media will be as commonplace as using the telephone, and probably much more widespread than watching TV. In most developed countries television consumption among young people is on the decline. Yes, they still spend lots of time in front of the screen, only it's a different screen: a computer monitor, namely. Recent studies reveal that in the United States, about five percent of all households no longer possess a TV set; not because they can't afford one, but because they don't need one.

And why should a sophisticated and autonomous youngster waste time watching prefabricated, commerce-driven and, let's be honest, for the most part deadly boring TV trash when behind the nearest computer screen a teeming, fascinating world awaits that has the added advantage of being completely tailored to my own tastes and preferences?

Throughout history the focus at every innovative step has been on the technology, so why should that be different with digital media? However, this time around the necessary amount of technical competency required to make best use of the latest innovation has been negligible from the very beginning. Time was when someone buying a car needed a crash course in auto mechanics to get by. Today nobody tinkers around under the hood unless he is a true car buff; the rest of us just want to turn on the ignition and drive away. And if for some reason the stupid car won't start, we grab the telephone, not a monkey wrench, and call up the nearest garage.

The authors of this book came of age to the Internet back when you had to know what a "TCP/IP stack" was. We both were very early users of CompuServe, a pioneering online service back in the 80ies and early 90ies, and we were sinfully proud of our five-digit ID numbers which marked us out as old hands. By the middle of the last decade, CompuServe was history and no one could remember how it was when you had to hook up a 56 Kbit modem, much less use an "acoustic coupler" which worked like a fax and sent bleeps and squeaks through a telephone line to be converted back into digital signals by the machine on the other end.

And all of this took less than 20 years! Today, online technology has faded into the background, and kids can connect wirelessly almost everywhere they go. The "ubiquitous Internet" we early users once dreamt about may not be fully here quite yet, but it's only a step away. And we are moving on already: In 10 years nobody will even say "I think I'll go online" – they'll always be online! They will carry around gadgets that may or may not resemble the smartphone of today; in fact they will probably look completely different, maybe like Google's Glass, maybe some kind of wearable "heads-up" display that follows us around and keeps shoving tiny snippets of digital information before our eyes, giving us an added digital dimension to reality. Who knows – and who cares?

As we become more and more comfortable in our digital surroundings and more adepts at dealing with the effects of digital acceleration, we ourselves will change. In fact, many of us already suffer from a new kind of restlessness. As Digital Natives, or at least Digital Immigrants, we are growing increasingly impatient since we have grown to expect more or less instant service on the Internet: Just press the button on your computer mouse and, voilà!, the results start streaming it.

Why stand in line for a movie ticket if you can buy it online? Why drive downtown to go to a book story when Amazon is just a click away? Especially since Amazon is investing heavily in "same day delivery", so the book I bought should be here in an hour or two. In 2012 alone, Amazon spent more than $550 million building 40 new distribution centers in North America alone. And it's not just books: Amazon will get a garden bench or a new mountain bike to you just as fast. Jeff Bezos, the founder of Amazon, after all never made a secret of the fact that he doesn't just want to be the biggest bookseller on the planet: he wants Amazon to be the biggest store in the world – period!

"Same-day delivery, far from being an exception, will be the new normal in a few years", readers of *Slate*, an online news magazine were recently informed.

The senior netizens among us can still remember the days when "WWW" meant "World Wide Wait", but those days are gone forever. Today, due to virtually limitless bandwidth, DSL and 3G connections (or even 4G; can 5G be far behind?), we enjoy truly instantaneous fulfillment of our every online wish. But while we may savor this as the culmination of our dreams, our kids think it's just normal – and even not fast enough. For them, Digital Acceleration is a fact of life.

Chapter 5

Generation Now!

Generation Now!

The pace set by digitalization and networking, the one that so many of us have problems keeping up with, has consequences for personality development, too. Common thinking has it that the best of our humanity is expressed through our individuality. This belief is especially popular among the right-wing Republican teapotties and with some short-sighted libertarians who hero-worship the individual, represented in its crudest form by tough guys like Marlboro Man.

Jaron Lanier, the former musician and Internet critic, is a typical example of this school of thought: Collective, or Swarm Intelligence, Crowdsourcing and the Open Source movement, he believes, are in fact harbingers of something he calls "digital Maoism" which will somehow stifle individual effort and lead one day to the downfall of the middle-class businessman. Been watching too many westerns lately, Jaron?

In fact, psychologists have long realized that individuation is simply one of many normal stages in the development of personality. Almost everyone goes through a phase of defiance at some time in their early lives. This is where adolescents learn to deal with conflict while exploring and developing their own distinctive talents, characteristics and limits. In most cases, this involves reacting against norms and values taught by others (parents, for instance) and establishing a moral compass of one's own. In the process, most of us learn the art of compromise and eventually come to grips with things we find disagreeable without despairing or going nuts.

The present generation (here the term is correct) is coming of age however without the experience of delayed satisfaction. Whether playing a computer game or buying something online, communicating with others or gratifying lust: Everything seems to happen in real time. Psychologists call this "instant gratification", and they believe it impedes other important stages in character development such the ability to resist the temptation for an immediate reward and wait for a later reward, also called "delayed gratification". Some psychologists link this ability to a wide range of other positive outcomes, including academic success, physical and mental health, as well as social competence.

So what happens is we no longer need to wait for our wishes to be granted and our desires fulfilled? What if waiting itself becomes unbearable? After all, the typical onliner no longer has to come to terms with the consequences of their own actions: All they need to do is press a key or click on the mouse to make things happen. And if not, the standard reaction of many is hot, blinding rage.

Digital bottle babies

We seem to be dealing with two lines of development here that are going on more or less simultaneously. The first is essentially progressive in nature: I am learning something new. The other is a kind of regressive infantilization: a step back in personality development – a return to the "original state" in the womb where it's warm and comfortable and everything is taken care of for us. It will be interesting to see how young people deal with the resulting psychological tensions.

Theoretically, they could turn into a generation of digital bottle babies. The question isn't whether this is good or bad. Instead we should be asking ourselves: Is it true or not? The process of infantilization is simply an implicit form of self-indulgence, and as such it has potential benefits. Through it, we may learn to deal with the pressure of living in real time. It may even help us develop vital new skills.

On the other hand, it could lead to frustration. After all, certain types of growing pains are an important step towards maturity. Without them, some may develop a tendency to sidestep frustration by avoiding certain tasks, thus failing to develop the social skills needed to enjoy and to play a role in social communities like family, policymaking or religion.

The widespread disillusionment with the political process among young voters is a good example for this. These kids grew up in a world full of opportunities for instant gratification, so they often lack the patience for protracted discussions about the issues and to search for necessary compromise. Political decision-making takes way too long for them, and there are other things to do in the meantime that are more fun and more self-gratifying.

In this situation, youngsters have exactly two choices: either turn your back on politics – or start a revolution! We saw how this works when the "Arab Spring" movement took off in late 2010, with pent-up demand for reform leading to a burst of "politics in real-time", something we will be discussing in more detail in chapter 9.

What was obvious from the start was that the results would be ambiguous: on the one hand "fast-track politics" made alternative forms of political dialogue suddenly look attractive, spawning the Occupy movement in America and the various "Pirate Party" in Europe. Suddenly politics was fun: you could be part of it and see change happening while you watched.

On the other hand, the minute politics starts to return to "business as normal" mode as it will once the dust has settled, the zest is lost, and the members of "Generation Now!", as we will call them from now on, turn back to their laptops, smartphones or game consoles.

So what are the alternatives? Which form of political organization can satisfy the coming generation's desire for instant gratification without becoming socially irrelevant?

Welcome to the Facebook Society

We believe that modern society needs to develop a new understanding of democracy. Digital Transformation and real-time decision making are empowering us and can pave the way a more direct, more personalized system. Instead of choosing representatives and sending them to Washington where they make our choices for us netizens should be allowed to choose for themselves. The outcome for society will be a system closer resembling online communities and social networks than yesterday's representative parliamentarian democracy. Decisions in such a society will be reached "on the fly" and through a variety of channels, depending on what decisions we need to make.

This is in fact just a return to the original idea on which democracy is based, namely small groups of informed citizens in the "polis" exchanging notions directly in the "agora", or marketplace. Imagine a lot of "digital agoras" where ideas and opinions are traded and where decisions are reached that effect only tiny communities which, in sum, are parts of the greater communality and finally of the world.

No wonder this scenario looks like a kind of "Facebook Society": That's what it is! People coming together as circles of "friends" (or maybe we should better say "online-acquaintances"), connected through digital conduits and able to both stay in contact and reach decisions for the common good.

Much will depend on how large social platforms like Facebook and Google+ develop as they begin to reach critical mass. If Facebook with its 1.4 billion members were a country, it would be as big as China and bigger than India! Anyone who belittles the importance of Social Media simply doesn't recognize the dynamic behind the development of what we will term the Online Society.

Platforms like Facebook and Google+ possess the decentralized patchwork structures needed to unleash a process of "re-politicizing" their users. Small groups within the larger system offer the potential for people to think and (inter)act as members of limited and manageable social environments with shared experiences and values. On Facebook I am never actually confronted with "everyone", although I can potentially reach and communicate with anyone I want. In fact I am usually just in touch with my fifty or at most a few hundred active "friends" (besides my thousand-odd less active ones) that have, through one means or another become part of my relational network, thus becoming my reference persons.

In such surroundings, interaction and decision-making happen significantly faster since I am only dealing with people who are important to me – and I for them. The resulting feeling of satisfaction is especially rewarding in an age of digitally limited impulse control, and it creates the desire to jump back into the teeming pool of participatory "grass-roots democracy".

In this context, infantilization can actually be a good thing: Kids have a clear advantage here compared to grown-ups who definitely worry too much!

The question, however, remains: What will the end result be? Will we all develop "Facebook Identities", similar to our present way of identifying ourselves as citizens of a nation, as adherents to some religious belief system or another, or as members of a community? Or will we remain members of tiny circles of friends and acquaintances to which we either belong for life or move around as the whim takes us, eschewing any really meaningful contact and without and real interchange with other groups within the broader community? And will my Facebook Identity replace other forms of cultural or spiritual self-awareness?

The answer is a resounding "yes and no!" National identity as an abstraction of ourselves will continue to lose significance. Kids in Europe today, at least outside the football stadium, see themselves less and less as "Germans", "Brits" or "Belgians". Instead, they think of themselves as "Bavarians", "Scots" or "Walloons", but even that is only a secondary identity. In reality, they see themselves as members of some community that convey identity on them, like my soccer club, my neighborhood, my circle of colleagues and friends – and yes, my Facebook friends, too.

The important thing is what interests me in my own particular place in life, what bothers me or makes me laugh (or click on the "thumbs-up button" on Facebook).

Speech and communication will continue to play a key role in this "Facebook Society", for the simple reason that speech is itself a form of identification. I feel at home where other people "speak my language", both literally and figuratively. On the other hand there is a clear tendency towards the development of a universal, world-wide "lingua franca" at least loosely based on English but spoken by kids in Bengal as well as in Bavaria or Birmingham. Hip-hop slang, an insider-dialect spoken around the world but often rubbished by professional linguists, is a perfect example: It may lack any discernable grammatical structure or established spelling, but hey, kids understand it!

Once again this just goes to show how human thought is influenced by our external perception of the present. What I am experiencing at this moment is real, always was real and will remain real forever. Other modes of identification such as language, music taste or cultural awareness can come and go. All you have to do is look at how the English language has evolved in the last 200 or 300 years to see that change is a constant and that human beings do a great job adapting to it. We, at least, don't worry ourselves unduly about the way language skills among Digital Natives are developing.

It's all mine, mine, mine!

By developing new sets of identities, individuals are placed in a position to set their own course and do their own thing. This process is often called "individuation" is that by which individuals differentiate themselves from other human beings, and it is a standard term in Jungian psychology. Digital networks reinforce this trend and allow us to invent our own rules. It's "my" style, "my" way of speaking, "my" way of expressing myself, and don't contradict me! In fact this is less a conscious decision and more a necessary part of character development. As individuals we are challenged to define ourselves in ways that make us stand out.

The "pop culture" of the 80ies and 90ies is a good place to start studying just how far the individuation of society has progressed. Only a few years ago it was possible for people to identify and describe themselves through "waves" or crazes that went under titles such as "beat generation" or "Rock'n'Rollers" and by their very nature transcended national or geographic borders. This is almost impossible to imagine any more. Today, the music world is subdivided into relatively small and close-knit "scenes" that develop concurrently and often overlap. Techno fans and New Romantics can share an evening out together, and Neoclassicists, Jazz fans or even New Hippies can and do live peacefully and in close proximity with each other. Each knows of the others, but it is no longer necessary to choose one or the other: It's everyone for himself and "chacun à son gout!"

Acceleration and individuation will have a big impact on how we satisfy our own needs and how we make lifestyle decisions in future. Increasingly the term "real-time" has been cropping up lately in discussions about consumers and retail. It's hardly a surprise that customers, too, are feeling the effects of digitalization.

Thus acceleration ceases to be an abstract notion and instead becomes part of our perceived reality. It makes a huge difference if I drive to the mall to buy a computer game or if I can simply download it. Anticipation used to be an essential part of the consumer experience; waiting for the mailman to deliver my purchase whetted our appetite and heightened our awareness. Now, the anticipation phase has shrunken to the few seconds it takes to pull dozens of megabytes off of a server half-way around the world. And game itself allows us to experience acceleration by telling us whether we've won or lost instantaneously, namely "in real time".

We can expect to experience acceleration in more and more parts of our lives, and we will become accustomed to measuring our relative satisfaction with suppliers of more and more products and services by the time it takes them to deliver results. Will the vendor make me happy fast, or do I need to sit around waiting for him to check my credit card details or process my order by hand before it can be completed? In that case, I think I'll go somewhere else!

Time killers or time savers?

While the digital world around us is changing at Internet speed, there are still those among us who are blissfully unaware of what's going on. Certain regions even in the developed world still lag far behind. Germans, for instance, are real digital slowpokes when it comes to using social media. Unlike the majority of Americans and Brits, at least 35 percent of all Germans refuse to use Facebook, as a study by ING Dipa Ipsos in 2014 revealed. One third of all Austrians mistrust social media in general. In Europe as a whole the rate is one in five.

Turkey, it turns out, is the leading country in Europe as far as social media usage goes: Half the population regularly uses Facebook, Twitter & Co, as became apparent in May 2014 when then premier Recep Erdogan threatened to shut down the servers at Twitter and YouTube because embarrassing videos about corruption in his government and even in his own family circle went viral.

Why are Europeans less enthusiastic about social media than Americans or, for instance, South Koreans where the number of Facebook users is expected to reach 13.6 million by 2016? A possible reason appears in a survey by Accountemps, a U.S. temporary employment agency, which found in late 2012 that 51 percent of the 1,400 top financial managers polled believed the productivity of their workers would decline if they allowed them to access their social media accounts during working hours. Many worried that their people could act "unprofessionally" on Facebook or Twitter, for instance by revealing company secrets or disparaging the competition. Around the same time an organization in the U.S. called Learnstuff published a report entitled "Social Media at Work" which claimed that social media was costing America $650 billion a year in lost productivity, though it does not detail where the productivity drop occurs.

The data may be soft, but the message itself caused lots of heated discussion, both online and off. Headlines proliferated in the old paper-based media, news anchors frowned. What does one expect: these people are engaged in a struggle for survival with the Web. And of course nobody reports on how much it costs the economy when employees read newspapers or watch TV on company time.

There are signs, however, that social media are in the process of mutating from time killers to time savers. And this would seem to be part of their growing up.

Facebook's business model, to name just one example, consists in luring users into their system and spending as much time there as possible. The longer a user stays on the platform, the more chances Facebook gets at displaying paid ads. This, incidentally, is something that analysts and investors worry about, because it requires users to be sitting in front of a big screen if the ads are supposed to work as they are intended. However, at least half of all users now routinely access the Web through mobile devices such as smartphones or iPads where screen size is limited. Facebook hasn't really found an answer to this problem yet, as anyone knows who has ever tried their mobile "app" which is, not to put too fine a point to it, pitiful.

Mobile users, it turns out, are very different, and this has to do with the circumstances under which they are using the Internet. Instead of leisurely "surfing" the Web, they are usually under pressure to find a certain piece of information "right now!": a telephone number to call, a price quote for a product they are standing in front of the supermarket shelf looking at, a train or plane arrival time or the weather forecast for this afternoon.

If this information is only displayed for instance if I watch an ad first, chances are I'll lose patience and go somewhere else. And besides, since I'm accessing the Internet through my phone or tablet, I'm in effect paying my mobile operator for the privilege of watching somebody's advertisement? No way!

Mobile marketing, just like the platforms that form the social web, needs to grow up fast if it wants to fulfill its promise of huge profits for businesses and operators. First-generation social media platforms are unsuited to this kind of use (and marketing strategies). They are essentially tied to the time-waster model: whether Twitter, YouTube or Pinterest, all need to keep their visitors captive in order to make money.

The next generation of applications for the social web are already on the horizon, and their model is very different: Instead of holding us hostage, they want to get us out and back to real life as quickly as possible. But before we go, they want us to stock up on some essential bit of information that will make life easier or more enjoyable.

Foursquare is the quintessential player in this new social media game. Thanks to a smartphone's ability to pinpoint our location, the system can provide helpful tips or user-generated information about, for instance, nearby restaurants or hotels, cultural attractions like museums or concert halls and of course promotional offers from vendors or service providers based on where I am at the moment.

Its twelve o'clock, so how about a nice pizza? Luigi's around the corner is offering a special discount for Foursquare users. Just click here and show your friends where you are and what you're doing (thus in effect making a recommendation), and will give you two toppings of your choice for free! Tell your friends that you have checked into our hotel and we will upgrade you to a junior suite!

The author and consultant Clay Shirky who also works as an assistant professor at New York University where he teaches media theory, believes that the next big thing in social media will be platforms that present themselves as extensions of and not alternatives to life in the real world. Instead of holding us up they will aim at enriching our lives with useful and finely targeted bits of information that are instantly useful.

Joichi Ito, the director of the MIT Media Lab in Boston, compares Facebook and other time-wasters to parasites as well as cast-off lovers. Both keep trying to insinuate themselves into our lives and feed off of us, in the first case by sucking our blood and in the second by whining in our ear. Both can detract or harm us, and in the end the parasite may kill us, while the former girlfriend will simply cause us to change our phone number or possibly move away. "Facebook tries to hook us", he says. "It wants to dominate our lives, just like a demanding lover or spouse."

At Foursquare, developers spend their time dreaming up new ways of reducing the time users have to spend on their platform, its founder Dennis Crowley once said in an interview. The goal is to limit the time of an average stay to less than 20 seconds. That's time enough for normal users to tell the world (or at least their friends and followers) where they are and what they're doing. Future social media apps will act as bridges between reality and virtuality; they will be like our toothbrush which is useful and makes us healthier but doesn't require us to think about or use it more than a couple of times a day.

No life without Facebook

Facebook, of course, belongs to the first generation of social web applications, and it will be interesting to see how they react to these new challenges. Will Facebook find a viable business model for the new mobile Internet? Word has been leaking out for a long time that their developers are working on it literally day and night, so far without very convincing results. Their success or failure will heavily influence Facebooks stock performance, and traders are keeping a wary eye on announcements coming out of their

Menlo Park headquarters. It was clear to most as far back as the initial public offering in March 2012 that Facebook was far behind in the mobile sector, and they have been struggling to catch up ever since. And Facebook is damned to succeed, if they don't want to disappear into obscurity almost as fast as they rose from their humble beginning in Harvard in 2004.

This would be painful for lot of kids and an increasing number of adults whose entire personal communication is based on Facebook. A young lady of our acquaintance who lives and works in Ireland recently told her parents that she would no longer write e-mails. "E-mail is so twentieth century, daddy", she explained. If they wanted to keep in touch, then mom and dad would just have to open an account at Facebook.

Young people in general, at least all over the developed world and in many emerging economies as well are virtually dependent on Facebook, and they can't imagine getting along without. But that is exactly what many would-be employers still are demanding from them. Too often during job interviews bosses still say: "Sorry, no Facebook during office hours."

A company that does that risks becoming "unworkforcable", so to speak: Kids will simply take their talents elsewhere. That could be fatal in an economy that is increasingly plagued by a shortage of skilled workers, no matter what age. Those talents we so desperately need to succeed in the future have different communication habits and needs, and cutting them off from their accustomed environment effectively makes them speechless --- and this almost worthless for the company employing them. After all, it was their instinctive ability to work and communicate through networks that made them so interesting for the company in the first place, wasn't it?

Thankfully, the dullards in the HR departments around the world are either approaching retirement age or learning to think twice. In 2009, *Der Spiegel*, a German news magazine, published an item in its online edition stating that 54 percent of all U.S. enterprises prohibited the use of social media during the working day. In 2012, a survey among large German companies conducted by Bitkom, the IT industry association, showed that only 30 percent persisted in forbidding Facebook and others.

The reason is clear: More and more companies are themselves present on Facebook where they are trying frantically to find ways of boosting their public images and improving their sympathy scores. So it's in their own interest to let their people use Facebook, Twitter or a host of other social apps to spread the word just how great it is to work for them. Companies today are in the strange position of needing to make their employees their friends (at least on Facebook).

Time was when blocking social media was considered a sensible precaution against data leakage and industrial espionage. Banks and other financial institutions were considered to be especially endangered, and with reason. Cybercriminals today often use methods described as "social engineering" to attack computer systems by getting individuals within the firm to let them in.

Facebook is an important tool for these crooks because it gives them names and faces of persons with important jobs inside their corporate victims whose computers might prove weakly defended against cyber-attack and which can then be used to dig deeper into the system, eventually reaching ones that are crucial to business operations or contain passwords and other information that can be profitably sold to competitors in China, Russia or elsewhere.

By sending friendly-sounding e-mails that appear to come from within the company, villains with good social engineering skills can often trick unwary users into handing them their passwords and other access information by posing for instance as colleagues in IT needing to do some necessary maintenance on their computers.

While there is a serious threat here, technologies exist to protect against them. Sadly, they are often not in place, due to ignorance or false economy. And at the same time, Facebook & Co. offer a chance to generate huge savings and/or revenue. It's not just all about funny pictures and piths Tweets: Deloitte, a consultancy, found in a study published in 1012 that Facebook generates 15 billion Euros a year in GDP growth in Europe alone. Small and medium-sized businesses stand to profit disproportionately from the chance to boost their presence and market products through the social web.

Everybody speaks for the company

Asking whether social media are time savers or time wasters becomes completely irrelevant at the point where we start talking about the so-called work-life balance, namely how do we separate our personal and our professional lives? The answer is simple: we don't! As we will be discussing in the next chapter, the borderlines are already almost indiscernible, and they will become more and more blurred as we move forward.

Whether we like it or not, we as employees of a company will be seen by our friends and acquaintances on the social web as representatives for the company we work for. As a result, everything we say on Facebook or Twitter will be seen in context with our relationship to our employer. Not only are our private and professional lives becoming one: our private and our business communications are coming together, too!

This has serious repercussions on corporate communications. In the old days before the social web, press speakers and marketing departments were responsible for the image a company projected to the world. Today, at least potentially, everyone in the company is a spokesperson.

Of course, press speaker is a delicate job at times and normally calls for a trained professional. Most regular employees never went to journalism school or trained in marketing communications. They will make serious bloomers and could possibly do the company great harm.

Assuming that large enterprises will not suddenly start sponsoring classes in public speaking for regular staff members, these employees will need to develop these skills on their own. Naturally, some will be better at it than others, just like everything else in life. And there are professional press people out there who suffer from foot-on-mouth disease, too.

The best thing would be to set out some ground rules. Many companies have taken to issuing so-called "social media guidelines". Unfortunately, many of them turn to their legal departments to do the job, and this is a sure recipe for disaster. The result is usually a twenty-page tome in small print that no one, not even a lawyer, will ever read. Secondly, the employees are doing most of their posting and tweeting after work on their own time, and a company will find it very hard to really enforce company guidelines after hours.

Managers need to understand that, instead of laying down the law, they will need to persuade and convince their people and get them to follow the rules voluntarily --- because they make sense!

Let us establish a rule of thumb here: Social media guidelines should fit on one side of a single sheet of paper and should not contain more than five or six rules! And ideally, they should be devised by the employees themselves. Simple: Call a staff meeting and tell them to please switch on their common sense for a couple of minutes. Ask them to tell you what they think in inappropriate behavior when talking about the company your work for on a blog or on Facebook. There aren't that many possible answers: Don't trash the competition, some of our customer may not like that. Don't reveal any business secrets because that could damage the company and its bottom line (and put your own job in jeopardy). Don't post pictures of your colleagues in an advanced state of inebriation at the last Christmas party because they might take a dim view, and besides, that could very well be a breach of privacy laws in your country of residence.

In many ways, this approach to setting the rules is similar to what we were all doing back in the early days of the Internet, when there was a big discussion about what constitutes proper "netiquette". Many of us wrote and published small rule books including things like "Respect other people's time and bandwidth", "Be forgiving of other people's mistakes", "don't start flame wars" (referring to especially violent and abusive exchanges of e-mails), and "adhere to the same standards of behavior online that you follow in real life."

And while these are all sound, sensible rules, maybe it would be better not to focus so much on what *not* to do and more on what we actually *want* to do. This harks back to our discussion about the new digital public in chapter 3. Neither companies nor employees can avoid public discussion about who they are and what they stand for anymore, so instead of trying to shut the discussion down they should greet it with willingness and zeal. It might be a good idea to remind readers here about the Boy Scout Motto: "Be prepared!" This means everybody in the company and not just a few trained individuals ordained as spokespeople. As we have been discussing all along in this book the responsibility lies with all of us, from big bosses to humble secretaries to use out powers of reason to "think the new!", in the process creating the necessary arguments and concepts on which to base a world of Digital Enlightenment.

Chapter 6

The New Life Plan

The New Life Plan

Critics of the Internet usually focus on digital acceleration and the transformation of our private and professional lives as something "unnatural", as well as on the apparent coming together of these two parts of our lives which used to be clearly isolated.

Transformation, especially if it involves fundamental change, can be frightening. In his historical novel "The Wine Dark Sea", Patrick O'Brian tells us that two Spaniards who are taking leave of each other will sometimes say "Que no hayan novedades", which can be roughly translated as "May no new thing arise" (until we meet again). That's just another way of saying that the new, by its very nature, is strange, unknown, and therefore potentially threatening.

What we experience personally through Digital Transformation is a sense of things speeding up until we get the feeling that we can't keep up anymore. There is even have a modern medical term for this, namely "burn-out" which was coined by Herbert Freudenberger, a German-born American psychologist, in 1974.

While use of the word has become widespread and "burnout syndrome" is a frequent diagnosis by doctors and therapists, it has also been described as a "myth", most notably by Yahoo-CEO Marissa Mayer as recently as 2012. "A lot of people work really hard for decades and decades, people like Winston Churchill and Einstein", Mayer said in an interview. Burnout, she maintains, is "about resentment. It's about knowing what matters to you so much that if you don't get it, that you're resentful."

We ourselves are skeptical. We have met or heard about scores of fully-charged and highly successful leaders in business and industry who have always operated at full speed; people like Steven Jobs or Larry Ellison who have probably never felt burned out in their entire lives. Those who feel overpowered are, of course, their subordinates, and understandably so. Keeping pace with a boss like Larry is surly quite exhausting. That doesn't mean, however, that middle managers have a natural right to ignore the dictates of Digital Acceleration and refuse to change with the world around them.

After work is so yesterday!

Fact is the networked worker is never "off". He or she continues to communicate with friends, colleagues and superiors without a glance at the clock and without strictly separating between home and office. As a true member of the emerging species "homo digitalis" he or she will write e-mails at any time of day or night and expect an answer "asap", regardless of time zone or office hours.

This growing sense of communication stress has led some to push back, for instance the worker's council at Volkswagen which forced management to agree in 2011 to shut down the firm's Blackberry servers half an hour after the end of the official working day so that employees couldn't answer e-mails from their bosses even if they wanted to.

Apparently, nobody noticed that a Blackberry is also a telephone, and that the boss could still ring his people up if he felt the need. However, let's all remember that every smartphone has a button you can press to turn it off completely...

The Franco-German IT company Atos has gone even a step further. Sending e-mails is completely prohibited, albeit for a completely different reason. CEO Thierry Breton decided back in February 2011 that he wanted Atos to become the first "Zero-E-Mail Company" in the world. In 1014, Breton announced "mission accomplished!"

In fact, Atos employees haven't foresworn e-mail completely, but the results in streamlining communications at Atos are still impressive. The company claims that so far it has saved 25% of work time previously spent on email and reduced disruptions and email overload by 60% — down from an average of 100 internal emails per employee per week to 40. What's more, over 200 business processes have already been certified as "email free", with more to come. Breton's aim is to transform Atos into a "social, collaborative enterprise where we share knowledge and find experts easily in order to respond to clients' needs quickly and efficiently, delivering tangible business results. First and foremost this requires a cultural change, learning new behaviors and management styles."

Most will probably agree that e-mail, something that almost everybody relies on to get their messages to friends, partners, co-workers and supervisors, is inefficient, to put it mildly. Mail can also induce harmful levels of stress as was demonstrated a couple of years ago by the Scottish computer scientist Karen Renaud of Glasgow University and the psychologist Judith Ramey of Paisley University. They asked 200 typical office workers how much time they spent looking in their electronic mailboxes. About a quarter of an hour a day, was the average answer. Then they installed a kind of snoopware of the test persons' computers and checked how often they really peeked inside. It turned out that the test persons were actually checking their mail an average of 40 times an hour! Which does beg the question: are we slaves to our e-mail programs?

Work without borders

Worldwide networks have virtually eliminated two factors that used to define our circles of endeavor, namely location and distance. In the age of digital networks it's not important where we are since we can be reached almost anywhere. And knowledge working is becoming more and more important, namely the ability to work with information and to transmit it to wherever it's needed whenever it's needed, the question of how far away I am becomes trivial. After all, the only noticeable difference lies in the network's latency which slows reaction time down by a few nanoseconds at the most, no matter where in the world I'm sitting right now.

Roya Maboob, a young Afghan woman, recently made *Time's* list of the "100 most influential people in the world" for her efforts at creating jobs for woman in her country, where for various social and religious reasons girls are often kept pent up at home, unable to attend school or to work for a living. Maboob's Afghan Citadel Software Company, an IT consulting firm founded in 2010, employs 25 people, 18 of whom are women who develop software for large international enterprises and even for NATO.

Because of threats from the Taliban they mostly work from home using the Internet to communicate and collaboration with their colleagues. Roya Maboob has since taken the idea a step further and has created 40 free Internet-enabled classrooms across Afghanistan to allow more than 160,000 female students to connect to the world outside.

Today, headlines often focus on the negative side effects of globalization as they are experienced by workers around the world, so maybe we should pause and take a look back to better understand how globalization came about and what it really means.

In 2005, author and columnist Thomas L. Friedmann described the three phases of globalization in his book "The World is Flat". According to him, globalization originally began in the 19th century with the invention of the steam engine and the rise of railroads. This first phase of globalization continued until about 1936, interrupted briefly by World War I. Then, Conrad Zuse's invention of the binary computer heralded in the phase Friedmann calls "Globalization 2.0" which is marked by the rapidly diminishing costs of communication and the rise of personal computers. The advent of the Internet gave this phase of globalization a huge boost.

Our current era, which Friedmann calls "Globalization 3.0" began around the turn of this century and is scheduled to last for a long time into the future. As more and more people gain access to potentially revolutionizing technologies that creates a "level playing field" with equal opportunities for all regardless of where they live or work. He identifies three main factors that are contributing to these developments:

1. Unlimited computer power that allows us to create and use potentially valuable content
2. Unlimited bandwidth which means we can send and receive valuable content anytime, anywhere
3. Unlimited collaboration thanks to powerful new tools and software systems that bring people everywhere together in a single digital working environment.

We might consider adding a fourth factor her, namely unlimited mobility. As mobile devices become smaller and more intelligent, allowing more and more of us to communicate and collaborate from anywhere in the world and around the clock, we are in the process of creating a new quality of intrapersonal exchange and communication which is radically reshaping how many of us knowledge workers earn our livings. We will eschew the hyperbolic habits of most media pundits by not calling this "globalization 4.0" but instead giving Tom Friedmann his due by sticking with "3.0".

By whatever name, this rapid diffusion of digital networks around the globe will have an overwhelming effect on the world of work as we know it. Prof. Wilhelm Bauer, the director of the Fraunhofer Institute of Labor Economics and Organization (IAO) in Stuttgart (Germany) describes this as what he calls "Work 2.0".

The problem is that most developed economies – and increasingly economies in the emerging markets as well – simply can't find enough qualified people to fill the rising demand for knowledge workers, a term coined by Peter Drucker in 1973 when he forecast that in twenty years' time it would be impossible for someone to keep up a middle class lifestyle by working with his hands.

A knowledge worker by definition is anyone involved in the tasks of planning, acquiring, searching, analyzing, organizing, storing, programming, distributing, marketing, or otherwise contributing to the transformation and commerce of information, as well as those (often the same people) who work at using the knowledge so produced.

As demand for this kind of worker escalates, many experts believe we will soon experience a "war for talent" on a global scale. And since our education systems are obviously doing a pretty poor job turning out enough well-qualified young people to keep up with the demands of business, the situation is approaching a crisis in many fields of business. A survey in 2012 among CEOs of large IT companies in the U.S. showed that 29 percent believe that their inability to find the talent they need is impacting strategic investments, causing them to either cancel or delay new initiatives. And 77 percent of CEOs are concerned about the unavailability of key skills, with 60 percent saying that this "talent gap" makes it harder to fill jobs.

According to Prof. Bauer, the solution lies in the "industrialization of knowledge work": A complex knowledge task, say a big presentation, an architectural design or a computer program can be separated into small modules that in turn can be farmed out to individuals who may not be fully qualified to do the entire job. Someone back in the company will have to assemble the results as they come in, but eventually the job will be done!

There are serious challenges here for the way businesses organize themselves. For instance: Why hire full-time employees if you can get results just as quickly (and probably much cheaper) by employing "freeters"? The term incidentally originated in Japan and describes young people who deliberately chose not to become salary-men, even though jobs are available. Freeters typically do not start a career after high school or university, but instead earn money from low skilled and low paid jobs.

This of course leads to another question: Will permanent jobs one day go the way of the Dodo? And will we all become freeters someday, and if so, who will safeguard the interests of the gainfully employed? What will the role of trade unions be in a networked world? Or will there no longer be trade unions because Facebook can do it much better? After all, should enough disgruntled knowledge workers get together and post complains about an unfair employer, won't that create just as much moral pressure as us all picketing in front of company headquarters?

The end of the full-time job will have consequences everywhere. Take office layouts, for instance. In the 60ies, partially-enclosed workspaces became popular: the "cubicle" was born. These were meant to revolutionize employee privacy and eliminate distractions. Today we typically see them as bland, isolating employee chambers for companies trapped in the past.

In recent years this has given way to something called "hot desks" or, as Prof. Bauer prefers to call them, "non-territorial workspaces." In effect this means that employees no longer have a desk of their own but instead sit down in the morning at the nearest free desk and work there until it's time to go home. The only condition is: you have to leave the desk as you found it.

"Clean desk policies" are commonplace today among IT companies and consultancies, and the notion is rapidly catching on in other branches as well. Studies show that about one-third of all employees in knowledge-intensive jobs are absent from their desks at any given moment: they're either sick or on leave or out visiting clients. And in many knowledge businesses working from the "home office" has become very popular. In fact it is often compulsory, at least for one or two days a week.

So will we one day all be free-ters (pardon the pun). For many, being self-employed and thus able to determine when (and increasingly where) we want to work is a highly attractive alternative to the nine-to-five drudgery required from us in the past, not to mention the daily commute from where we choose to live to where we are forced to work.

As education levels rise and many of us explore alternate lifestyle choices, being freed from full-time employment may appear as a real blessing. And most of the digital tools we will need to work as, when and where we wish are already here.

Work 2.0 and other forms of job flexibility are actually good for most people as they provide a new degree of freedom to shape our own working environment as we see fit. Instead of being "office slaves" we are offered the opportunity for more self-determination – a necessary precondition for digital autonomy, which as we will see is an important prerequisite for Digital Enlightenment.

Of course, many feel uncomfortable with the thought of being set loose, for instance being required to self-discipline themselves and keep their productivity levels up. Home offices offer a multitude of distractions, and many are unaccustomed to setting their own pace and their own goals. There are a number of answers to these fears, but the fact remains that some will simply not be able to cope and will therefore be in danger of falling into a kind of mental and physical atrophy if no one is there to keep them motivated.

Like other forms of human handicap, it will be society's obligation to provide assistance for these people. However, the rest of us can't simply wait for these unfortunates to catch up. And no one can turn back the wheel of time.

Besides, if you still want to go to the office – why not? That, too, will be a choice for most of us. And after all, even if we choose not to remain tethered to our desks, there will still be plenty of things to do at the office. Brainstorming, for instance, works better, at least in our own experience, if you can look the others in the eye. And there are lots of opportunities to party, for instance to celebrate a successful business deal or a job well done. But go to the office simply to work? Who wants that?

Digitalization versus industrialization

The new world of knowledge work described above deals with information and ideas, and the degree of Digital Transformation going on there is breathtaking. However it is by no means restricted to the more administrative or creative professions, but will be felt just as keenly in such "old economy" fields as production and industry. Here we are dealing with more tangible things, namely manufactured goods. In April 2013, the British business magazine *The Economist* came out with a special section dealing with what they called the "third industrial revolution. German economist and politicians, on the other hand, routinely refer to something called "Industrie 4.0", a term enshrined in official government documents dealing with a program entitled *Hightech Strategie 2020 for Germany*.

Again, there seems to be some confusion as to how we number the succeeding generations of development, but at least virtually all experts agree that the first great Industrial Revolution took off in England in the late 18th century where water and steam power became widely available. This led to a switch from hand production methods to the widespread use of machines and the development of machine tools as well as the change from wood to coal.

Differences of opinion already start to appear as to whether or not the so-called Second Industrial Revolution, which is thought to have been led by developments in the latter half of the 19th century until World War I, mainly led by America, are not in fact simply an extension of the first Industrial Revolution or a separate epoch. In any case, it was characterized by rapid expansion of railroads, large scale iron and steel production, widespread use of machinery in manufacturing, the use of oil and the electricity as well as first primitive forms of electrical communications.

The third revolution described by the *Economist* is predicated and driven by PCs and the Internet, and we are right in the middle of it today. The interesting question is whether the next stage in the development of the Internet, often called the "Internet of Things", should simply be seen as the logical extension of the existing Internet, or whether it deserves a separate designation as the Fourth Industrial Revolution.

Again, we choose to stay closer down to earth, so we will go with our friends at the *Economist* and only count to three. However, we do believe that the term "industrial" revolution is a slight misnomer. In fact what is happening is the exact reverse of an industrial revolution as it will infallibly lead to large-scale "de-industrialization" and a return to craftsmanship and manufacture in the true sense of the word which comes from the Middle Latin "manu" for hand and "factura" for the working of metal and other materials.

Just as digital networks have expedited and democratized the exchange of information (everyone can be at the same time reader, author and publisher at the same time!), so too is it at least theoretically possible for each of to design and build everyday objects or to cause them to be built to our own specifications by companies versed in the burgeoning art of "individual mass-production."

This isn't as new as it may sound. The automobile industry has been doing it now for years: Buyers access an application often called the "car configurator" on the company's website where they are offered hundreds or thousands of choices, ranging from different engine sizes to your choice car color and interior trim (leather bucket seats, anyone?) In an interview, Dr. Rainer Feurer, head of corporate strategy at BMW in Munich, recently told one of the authors that "we never see two identical cars leaving the production line one after another. We only build one-offs nowadays."

At the annual trade show "Euromold" in Frankfurt visitors can catch a glimpse of the future today as vendors from around the world showcase their 3D printers and other forms of industrial production equipment aimed at allowing ever greater degrees of variation at increasingly low cost. Gone are the days of giant machines covered in oil and grit and clanking loudly as they crank out identical instances of some industrial product or another. Today, anything from running shoes to sunglasses, from electronic components to automotive parts, from dental implants to mobile phone covers can be cranked out almost instantly by copying a digital template and extruding plastics or metals through the tiny nozzles of 3D printers.

In December 2012, a group of so-called "hacktivists" going under the name "Defense Distributed" demonstrated their ability to build a fully-functioning pistol out of heat-resistant plastic, using a store-bought 3D printer, subsequently conducting shooting tests with live ammunition. When they published the digital master plan on their homepage, the Department of State stepped in an shut down the website. Not that this stopped the plans and instructions from proliferating widely through proxy sites. Note to self: On the Internet, information wants to be free, dude!

3D printers stand the principle of industrial production on its head. If you were to ask a factory owner to build you a single hammer, he might possibly be able to do it, but it would sure cost you! He would have to create a mold, melt the iron or steel, cast a work piece, have somebody polish it, go into the woods and find a good-looking branch, trim it down and make a shaft, and so on. This is utterly uneconomical, of course. To make a profit, the factory owner needs to make thousands of hammers; economies of scale take care of the rest.

The 3D printer suspends the laws of scale economy. The software can be copies infinitely and modified quickly and cheaply. Setup costs are the same every time he makes a hammer, and the process can be repeated as often as desired by simply pressing a button – at least until it's time to change the printer cartridge.

Of course factories won't all have to close just because of the invention of 3D printers. But industrial mass production will experience radical change due to digitalization and networking, too. The results will be just as disruptive as those we have seen in other areas of the economy, like the music or film industries, publishing, book dealers, photographers and many, many more.

In Schenectady, New York, General Electric has built a factory that manufacturers batteries, and being typically American they flamboyantly call it the "factory of tomorrow". The plant is laced with sensors gathering data on just about everything, from the temperatures at various points within the ovens to the amount of energy being expended to produce any given battery, not to mention the humidity in the plant's vicinity which can also influence the final quality of the product. As humidity rises vents in the air conditioning open admitting less moist air into the ventilation system.

But General Electric goes a big step further by making the end users themselves part of the manufacturing process. Batteries made here are used in cars and large machines which are also fitted with lots of sensors which are connected to the Internet via Wireless LAN and are capable of sending performance data back to the factory floor, indicating for instance when and what kind of problem may have occurred. That means that GE can warn its customers that a battery may be about to fail and actually send out a maintenance technician to fix or exchange the battery even before anything goes wrong.

Jody Markopoulos, CEO of GE's Intelligent Platforms division. Believes that industrial production will come under huge pressure because of global networks. Competition will increase because customers will demand more and more individualized products and the pace of technological development will become even faster. In such a volatile environment, Markopoulos thinks, manufacturers will be forced to bring improved products to market at ever shorter intervals and at less cost.

Only through extensive use of digital networks will they be able to stay the course, she says. "We need to make physical manufacture more intelligent by connecting our production equipment directly to the Internet. That way, we can combine data from the assembly line with market data, gathering new insights about how we are doing our job and how our products perform, thereby enabling us to improve quality as we go along which will in turn help us sell more."

Markopoulos uses the term "Industrial Big Data" to describe this process, and she believes it is already heavily influencing the world of industrial production. The focus is on gathering as much information as possible about people, processes and production facilities and evaluating them in real time. Software will be used to discover hidden trends in the masses of data and to make intelligent guesses, for instance about when and why machinery could break down and how variations in quality occur. Foremen will walk across the shop floor equipped with smartphones or iPads that will display instant warnings the second something unusual happens, and they will be able to fix most problems remotely – just like computer system administrators now do every day.

"The role of industrial production is shifting", a group of McKinsey analysts wrote in autumn of 2012 in a report on industrial globalization. While manufacturing used to contribute to GDP through growth and job creation, in future things like innovation, improved productivity and exchange of goods will become more important. Manufacturers will take advantage of new services and opportunities to outsource parts of their production, making them increasingly dependent on global network effects.

This is apparent even today as large industrial concerns such as car makers, pharmaceuticals or clothing manufacturers increase their efforts to create "global production platforms": factories that can be retooled on the fly to react immediately to changes in market demand. In 2012, Volkswagen introduced a concept it calls "MQB" which stands for „Modularer Querbaukasten", or "Modular Transverse Matrix", and which allows the company to install identical units in a total of 43 models from their Audi, Seat, Škoda, and Volkswagen brands of cars and one day soon into trucks as well. The system has been installed at every Volkswagen plant, from Wolfsburg in Germany to Uitenhage in South Africa, from Changchun in China to Chattanooga, Tennessee. The result is not only a weight saving of 60 kilograms per car, but especially an unheard of degree of flexibility in line production.

The old way of making stuff, let's call it "Industry 1.0", is on its way out, but that's not all. Its economic importance is shrinking, too. Manufacturing used to account for 25 percent of GDP in Germany in 1975; by 2010 it had fallen to 15 percent.

The future belongs to "digital manufacturing". This means a world in which small groups or even individuals will be able to design and produce everyday necessities themselves, faster and cheaper than any big factory. New materials and new production methods such as 3D printer's mentions above will proliferate and change almost everything. On the technological horizon we can already see things like personal robots or concepts such as "collaborative manufacturing". A good example is a company called Quirky, a tiny design studio located in an old warehouse on the banks of the Hudson in New York where product ideas submitted via Internet from consumers around the country are transformed into prototypes using an extensive range of laser cutters, 3D printers, conventional lathes, and modern CNC equipment.

Quirky not only builds products to order, it also helps its customers set up shop as vendors of their own products. The most successful project at the time of writing was something called "Pivot Power", a flexible multiple socket outlet capable of holding even the biggest and most clumsy power adapters. The inventor, not surprisingly, was a computer geek, Jake Zien of Milwaukee, who was fed up juggling with all the adapters for his many gadgets. By late 2012 he had already sold more than 200,000 of his brightly colored bendable power strips.

Quirky had taken care of everything from making the actual prototype to registering a patent and convincing a bank to lend him the necessary capital. And through its online community Quirky was able to gather feedback and testimonials from actual customers in very short time, plus getting news of the new product out to the market almost overnight.

Ben Kaufman, the founder of Quirky, told *Forbes* that he is convinced that collaborative manufacturing will be a major trend because it allows the collective fantasy of the crowd to transform itself into real, tangible products. "A machine like a 3D printer can make anything you can imagine, he says --- and many you can't, at least not yet.

Digital Bedouin seeks digital oasis

There is no avoiding the fact that digital networks will change the way we work. The question that remains, however, is this: How will we cope? After all, we can't all hope to profit equally; there are always winners and losers.

The Industrial Revolution, says the British economist Dr. Leigh Shaw-Taylor of Cambridge, was the direct result of the Enclosure Movement in 18th and 19th century England: villagers lost their traditional rights such as mowing meadows for hay, or grazing livestock on common land formerly held in the open field system and were forced to move to towns and cities where the increased labor supply contributed to the rapid growth of factories.

At the same time, the growing town populations increased demand for manufactured goods: a cycle that was perceived by its victims as vicious, even if economically speaking the benefits for society as a whole were obvious. At least it gave Great Britain a head start in industrializing that lasted until long into the 20th century.

Like enclosure, digitalization and networking are disruptive technologies that threaten to uproot many of us working in the most effected professions and therefore capable of triggering existential fears in us. Digital acceleration contributes to this confusion, as did the World Wide Web which was arguably the most disruptive invention in history, both for the individual and for society.

Individual mobility takes this even further. As we move around with our smartphones and our iPads at the ready, always on and always reachable, we discover a strange paradox: Why travel when we could just as well stay at home and write e-mails? If videophones are capable of transmitting images so crystal-clear that we can see every tiny hair in the other guy's nose, then why aren't the number of business trips taken worldwide declining?

In fact they're going up all the time, as IATA, the international airline association, confirms in its latest growth forecast which indicates that passenger flights are set to increase by an average of 5.4 percent annually. People, it seems, are more on the move than

ever – and they take their work with them.

Every morning the lounges at large airports all over the world look completely alike: yawning businessmen sitting at tables, coffee cup in hand, silently staring at the screens of their laptop computers or tablet PCs, checking e-mail on the iPhones or talking excitedly with colleagues who may be sitting just down the street or half a world away with whom they may very well meeting in a few hours to discuss the latest deal or project.

These are always the same people, and they form a kind of social subset. What to call them? "Teleworker" was once popular, but today it conjures up an image of people wearing clumsy headsets crammed into cubicles at some call center. And "home worker" doesn't do it, either, does it? These people work wherever they happen to be, in the airport, at in bars and restaurants or even on a park bench.

In America, the term "road warrior" used to be popular, but it doesn't really describe what's going on very well. After all, most of these "warriors" aren't all that ferocious...

Nicholas Negroponte, the legendary head of MIT Media Lab in Boston, once coined the phrase "digital nomad" which he used to describe someone who carries everything around he or she needs to be productive. Unfortunately, that usually includes a load of gadgets, adapters, sticks, drives and chargers as well as files and books that often fill more than one briefcase or even a small suitcase. Who wants to be a nomad?

Possibly the best label devised for these mobile workers appeared in an article published by *The Economist* as far back as 2008 under the headline "The Digital Bedouins". Unlike former teleworkers who resembled astronauts – people who carry their own oxygen around with them – Bedouins follow a tribal lifestyle that enables them to travel unencumbered, essentially equipped only with a burnouse, a camel and a small bottle of water. And they don't need anything else to survive since they know exactly where they will find the nearest oasis.

The digital Bedouin's situation is similar: all he needs is a smartphone or tablet; his oasis could be the nearest Starbuck's where he can find all he needs to survive in a digital world: an Internet hotspot! The coffee is just a perk (or a pain, depending on your taste). We personally mostly chose to purchase a latte just to show our appreciation; after all, you can't sit around for hours pulling mail and not consume something, can you?

Our blogger friend and columnist for the newsmagazine *Der Spiegel*, Sascha Lobo, in his book "They call it Work", describes the unconventional lifestyle cultivated by young people in Berlin, one if the most wired cities in Europe, whom he calls the "Digital Boehme". These people, he says, sit around all day in coffee houses, just like their ancestors in Paris or Milan, but instead of books and writing pads they bring their laptops with them. Nothing else, though, since that's all they need to communicate, socialize and earn a living. Between fits of inspiration or industry they read newspapers, chat with friend or head for the door, calling back over their shoulders: "Gotta go work – my landlord wants some rent."

This new form of liberty changes many things, not the least those concerned. It seems almost like a new species has been born: "homo mobilis", the mobile human. Mobility not only alters our behavior, it also changes our self-awareness. By heaving off the shackles of time and location we become different creatures, unlike our more stationary (or should we say: sedentary) fellow-humans.

And while stationary workers produce more and more paper, techno Bedouins don't print things out, they store them in the Cloud where they are instantly available – and sharable – anytime. At the same time, digital Bedouins are less aware of the technology they use because it isn't that important: hey, it's just there, and it works! Anyone can operate an iPhone without reading a manual first. And so the final obstacle is removed on our way to true, total mobility.

Of course, digital Bedouins don't have time clocks. The working day has 24 hours, and anyone who is self-employed has a tyrant for a boss, says our friend Paul Saffo from the Institute for the Future in Palo Alto. This can cause problems at home ("no Blackberry in bed!") and can lead to overload if mobile man doesn't remember to switch off once in a while. Pity the so-called "crackberries"; people gripped by a terrible compulsion to constantly check their Blackberry devices. These are the real victims of Digital Transformation, and they need our help!

Thankfully, most members of "homo mobilis" are learning the important lesson that more information doesn't necessarily lead to more intelligent decisions. The techno Bedouin realizes that the ability to filter information, and to only store things he really needs in his biological memory, is the most important lesson we all can learn. Everything else we can find online.

Chapter 7

Welcome to the Global Village!

Welcome to the Global Village!

Dr. Emilio Mordini is a witty man, and an extremely intelligent one, too. He's an Italian, a psychoanalyst and a computer scientist (a dangerous combination if ever there was one), so he knows men's souls just as well as he knows their computers. In Munich in 2012 he gave a speech entitled "Secrets and Privacy in the Age of WikiLeaks" in which he described the total transparency of the digital era and the consequences for society and for our mental health.

Secrets, Mordini maintained, are important if we all want to remain mentally stable. If we thought everybody around us knew all our inner secrets we might actually go crazy. We want our feelings, longings and habits to remain hush-hush because it makes us feel comfortable and well-balanced. But of course, he added, nothing is really secret, at least not for long, and it's getting harder and harder to keep something secret as we move deeper and deeper into the digital age. But on the other hand, he reassured his audience, candor is actually mankind's natural state. After all, our species spent most of its evolution living in village societies; and in a village, nothing is secret.

"In the village, everyone knows everything about everyone else", Mordini said in the warm, reassuring tone of a psychiatrist speaking to a patient on the couch, "but of course everybody acts as though it were a secret. That's important if life in the village is supposed to go on as normal. If anybody blabbed, the peace of the village would soon be shattered."

Pulcinella's Secret and the invention of privacy

Mordini isn't the first one to describe society today to a "global village". Marshall McLuhan, the father of modern communication theory ("the medium is the message") did so in the early 60ies. But Mordini takes the analogy further. It seems that Italians are familiar with something they call "it segreto di Pulchinella", or "Pulchinella's Secret". It's from the traditional folk theater called "Comedia dell àrte" that comes down to us from the 16th century Renaissance and is still popular today throughout the Apennine Peninsula where it is performed in small-town theaters, halls and even tents.

In one of the *lazzi* or comic dialogues someone tells Pulchinella, who is sort of the village idiot, a secret but admonishes him not to tell it to anyone. Poor Pulchinella, of course, is incapable of keeping a secret, so he goes around and tells one actor after the other what the secret is, but always adds with a finger to his lips: "Psst, it's secret!"

Soon, everyone in the cast and in the audience are completely familiar with the secret, but of course the piece goes on. Things become very complicated and funny, but only in the end does everybody admit that they, too, are party to the secret.

Life today is full of "segreto di Pulchinella", Mordini believes. Take computer security, for instance. Companies spend small fortunes (and sometimes not so small ones either) in securing their networks and servers. But ask a IT security expert, and if he's honest he will tell you that in really is no such thing as a secure computer. At least not as long as its connected to the Internet (or even if it's turned on, for that matter). Our friend Kim Cameron, the head of Identity & Access Management at Microsoft in Seattle, once postulated that "sensitive information will be leaked", a statement that has since become known in security circles as "Kim's Law". How can the best firewall in the world help you if a disgruntled employee, perhaps even a trusted system administrator, decides to leave and take your most sensitive data with him? And what if someone decides it's his duty to "liberate" your data by hacking into your system (or physically breaking in and stealing your computer) so that the results end up on WikiLeaks?

Nothing can be secret, Mordini says, because if it really were, than the information would be useless. A secret, by definition, remains restricted to a limited circle. If the last member of the circle dies, the secret remains, only it has become irrelevant. Secrets, Mordini says, are artefacts we construct in order to retain our sanity.

What happened in the barn

From time to time, we read in the news about someone running suddenly amok, and often it turns out that the reason was the culprit believed someone else had found him out. We call these people "obsessed", but of course, just because I'm paranoid don't mean they aren't out to get me, as the old saying goes.

In 1913, a school teacher named Ernst August Wagner got up one morning, murdered his wife and four children and took a train to the nearby village of Muhlhausen, where he used to live. There he waited until after dark, then set fire to a barn and waited for the villagers to appear attempting to put out the fire, when he quite calmly started shooting them as they ran around silhouetted by the flames. Twelve people died, and Wagner became the first convicted criminal in German history to be locked away in an insane asylum for life instead of being beheaded, as was then the normal form of capital punishment.

The point Mordini tried to make in his speech was this: If we have the feeling our deepest secrets aren't secret anymore, we can become seriously destabilized or even cross the border into insanity. The question is: what is cause and what is effect?

Going public

To understand this, we need to remind ourselves that privacy and secrecy are not the normal human condition, as stated above. In fact, privacy is an invention of the 18th and 19th centuries. The rising middle-classes suddenly found themselves wealthy enough to build private houses big enough for them to "retire into the privacy of the home", as the adage goes. Formerly this had only been the privilege of the very rich or of the aristocracy.

The notion of privacy as more than a luxury for the very few, therefore, is less than 200 years old. The rest of human history was spent in village culture, where everything is known about everybody and secrets are shared but hardly ever mentioned.

In fact, it turns out that privacy is actually a very Western concept. In India, even middle-class people are used to living in large, densely packed family units where privacy is almost completely unknown. In Japan, where the walls of rooms in the average house consist of rice paper and bamboo, any attempt to "retire into privacy" would be though rather funny. In most Eastern societies, everyone knows and accepts that they can, at any time, become the instant focus of a large gathering of family and friends, neighbors and strangers who may wander in from the street.

The concept of an "open society", then, is already an accepted norm in many countries of the world. Anyone who is part of such a society lives with a kind of moral reflex that says: Be aware that anything you do or say is probably already part of the public domain or can become so at a moment's notice. Live your life accordingly.

Japan is a good practical example of this principle at work. There, individuals are hardly ever really alone, and over time society has developed certain rules and habits that assure a certain amount of discretion. For instance, if you live side by side, separated only by a paper wall, others will presumably hear every sound you make. Young Japanese girls are typically very bashful, so many of them feel embarrassed by the sounds they make when going to the toilette. In order to drown them out, many start flushing as soon as they sit down, comfortable in the knowledge that unwitting listeners can't distinguish between the sounds being produced naturally, so to speak, from those emanating from the plumbing.

On the other hand, young Japanese belong to one of the most environmentally conscious generations anywhere, so many of these girls worry about wasting precious drinking water this way. Which in turn has led business-savvy entrepreneurs to jump at the opportunity by offering them gadgets that produce the sound of running water electronically! One of the most famous is called "Otohime", or "Sound Princess".

Actually, Otohime is a mythical Japanese princess whose father is the God of the Sea. Anyway, these gadgets can be found in women's WCs all over Japan. Funnily, they never caught on with men, who apparently could care less who's listening...

In the United States there is a movement that calls itself "Post-Privacy". Like many techno-libertarians before them such as the eternal hippie and *Grateful Dead* songwriter John Perry Barlow these people postulate a "Human right to information" and demand that confidentiality and data protection be outlawed as a crime against humanity. This includes, of course, such things as information about social contacts, political leanings, personal outlooks, financial status and health records. This applies not only to individuals however, but also to government.

Nova Spivack (which may or may not be the author's true name) wrote in an article for *wired* in 2013: "Even the CIA and NSA cannot protect their own secrets as effectively as before. The same holds for corporations". Increasing transparency levels would level the playing field, he (or she) maintained. "Everyone can watch everyone. Whistleblowers are everywhere; it's mutually-assured disclosure."

Instead of trying to hide secrets, we should focus our attention on how to share them, the author believes, concluding that "Healthy transparency is not the opposite of privacy. In fact to be sustainable, transparency is respect for privacy in practice, even when it does not actually exist in principle."

If you don't want something to become public, then don't put it on the Internet, one might add. Or even on a computer.

"For classic data security people this is a nightmare", writes Stefan Münz, a German Internet activist and politician in his blog, adding that

"Most politicians are confused, and rightly so. Power users are busy posting their motion profiles openly and negligently on websites like Gowalla or Foursquare and then turning around and complaining on Twitter and blogs about state-sponsored snooping programs like the European Data Retention Policy. Sounds like a contradiction, but it's actually pretty simple: these people want a state that suits their way of life: transparent, open and ready to share all it knows about us openly. Among themselves they're used to this: Open Source, cooperation, putting your cards down. This is no fuzzy dream of Western-style socialism we used to talk about in smoke-filled student pubs long after midnight; this is real life! Not goose-step mentality disguised as anti-capitalism but a laid-back, laissez faire attitude within the networked community that accepts egotism as long as its conforms to their notions of what is acceptable, but will combat egotism that it perceives as against the interest of the group; these are punished by digital exile."

Digital Omerta

There is a big difference between believing in real "privacy" as an area in which our private lives are conducted and to which no one has unauthorized access, or whether we say, okay since everything is public anyway, let's agree on some rules governing discretion. These are in fact commonplace. Just think of the unspoken law of "Omerta" which prohibits members of the mafia and their families in the mountain villages of Sicily from speaking to outsiders about "private" issues.

Pulchinella's Secret and the principle of Digital Discretion take this a step further by asking us to act as if we hadn't seen, read or heard anything. This is essentially a gesture of consideration for others by allowing them to think something is secret when in fact it isn't – and can't be. At least we won't talk about it, how's that? Thus a kind of nice fiction is created which everyone can accept and live with; even though if you think about it for a while you realize it is in fact a fiction.

Discretion is a rule of behavior, not an abstract concept like privacy. The so-called realm of privacy is simply an idea with a certain historical background, and it may have been a good idea at the time of its development but now it is rapidly becoming outdated. Imagining that there is somehow a place you can retreat to and enjoy "my privacy" is about as antiquated in the digital age as insisting on writing with a quill. It's simply obsolete.

The question remains: How does the perception of people change in this situation, and does this lead to a willingness to accept and play by new rules? We stand today at a divide: on one side those groups and individuals who believe we need to do much more to protect our precious privacy, especially in the digital sphere, on the other there are those who thinks this is all just a huge waste of time, so let's make "public" our default setting.

On Facebook you can see both sides in action. Many Facebook users do everything in front on an audience which can include almost everybody. These are the limelight hogs of the digital age for whom Facebook is simply a big stage where they can present themselves with various props and in different poses. For others, the so-called privacy settings on Facebook are important because they allow them to manage who gets to see what they're doing and posting, not realizing or caring that Facebook reserves its right to store everything and use it more or less as they see fit. And finally there is the tiny and diminishing group who believe that Facebook is devil's work anyway.

We believe there is a growing town country gap here as well as a growing generation gap. The latter however is less clearly marked: Young people tend as a rule to open up on Facebook more frequently than their elders, but exceptions often prove the rule. Common warnings by the older generation include admonitions not to do anything on Facebook that could be used against them by the boss when youngsters apply for a job, but often it's the youngster themselves who caution their peers to be more circumspect when using the social web.

The result is a more differentiated view of Facebook and its users. A young lady of our acquaintance recently refused to publish pictures of her new-born twins on Facebook, where she is otherwise quite active. When questioned she hemmed and hawed at first and then burst out: "They should be big enough to decide for themselves first."

A look at typical samples of use patterns by young people on Facebook reveals a number of opposing trends, for instance:

Limited exposure: Many users no longer seek to grow their circle of friends almost frantically as a way to show off and gain status. Instead, they keep a low profile so as to be able to handle individual conversations as they would with "real" friends. Typical quote: "I don't want more than 160 Facebook friends. I don't really know more people than that!" (Felix, 20, sociology student)

Facebook eruptions: Posting too frequently and on too wide a range of subjects is considered to be "uncool" by many young users. "Don't befriend that girl, you can't read all she posts anyway!" (Annika, 22, studying to become a teacher)

Private parties: Invitations are only sent out to a select group versus simply asking everyone to join the fun (which often leads to headlines about hordes of strangers crashing Facebook parties and creating mayhem). Invitees are expressly asked to refrain from posting pictures of the host or other guests without express permission. (Heike, 24, accountant)

The contradictions here are clearly apparent: Neither the fans of radical publicity nor the anxious and implacable believers in strong data and privacy protection laws will conceivably win out in the end. Instead, the so-called digital natives themselves are the ones who appear to be putting the idea of digital discretion into practice most convincingly. "If I'm talking with my friends on Facebook and my boss is snooping, then he's the asshole, not me", a young friend of ours recently maintained when questioned. Digital peeping toms are increasingly being seen by the young as acting indecently and embarrassing themselves.

So in the end Facebook users will divide themselves into disparate groups, on the one hand those who can't post enough pictures of tiny Tim online and on the other those who, while probably just as proud of their offspring, prefer to keep them out of the public eye.

The Rumpelstiltskin effect

It is worth noting here that, in India, children are hardly ever called by their real names until they grow older. Instead, everyone in the family and circle of friends uses a nickname chosen by common consent. This is supposed to protect kids from demons and other evil spirits by confusing and distracting them away from their victims.

On the Internet, this method of disguising one's true identity by adopting a nickname, or "handle, is actually quite ordinary and is valued especially by some older users as a form of self-liberation. And frankly, it can be fun to change one's identity or even one's gender at will online, kind of like attending a fancy dress party where everyone wears a mask and no one is who he or she seems to be. Studies show, by the way, that women tend to practice "gender switching" more often than men who surf the web.

We call this the "Rumpelstiltskin Effect", and it is more common than you would think because it gives us a way to live out some of our most intimate fantasies, kind of like starring in an online Walter Mitty movie. Of course, sometimes the fantasies go too far and the police come crashing through the door to arrest the respectable family man who's been spending the late-night hours illicitly downloading kiddie porn onto his hard drive.

But don't let that stop us from playing hide-and-seek online for as long as we want to. Even the most avid libertarian radicals will grant that anyone can do whatever he or she wants on the Internet as long as we don't harm others. By "harm" we mean demeaning someone or infringing on their freedoms of expression and opinion. Which brings us to the emotionally charged issue of online censorship: Sex and violence, both part and parcel of Western culture, have proliferated at an alarming rate in the land behind the computer screen, and this worries some, especially some politicians.

In Germany, for instance, the then minister for family affairs, Ursula von der Leyen, gained a nickname as "Censorship Uschi" for her suggestion that the authorities take down websites offering child pornography and instead force Internet providers to display a bright red stop sign. For this she caused an association to be founded which was quickly joined by other prominent politicians from her center-right Christion Democratic party, was well industry groups and technology companies, all calling for a "clean Internet".

The initiative was launched with great fanfare but soon sort of petered out when it was found that most of the questionable websites weren't being hosted in Germany, where Ms. Von der Leyen could conceivable gotten hold of the perpetrators, but somewhere in Russia or the Caribbean where she couldn't.

About a year later, the following press release was sent out via e-mail: "The move by the federal government for blocking websites has shown that 'going it alone' often works only selectively and that in particular purely technical measures can only indirectly help solving the problem."

This, of course, is politico-speak for: "Sorry, we blew it!"

Under the digital veil

There are two additional phenomena often discussed under the heading of privacy and discretion, namely anonymity and pseudonymity. Young people growing up in a digital world soon learn all about these two ways of hiding one's identity online, and for them they are elements of what we might call the "new privacy".

In a world where everything is open and transparent, anonymity can be a place of retreat, almost a kind of digital veil we pull before our faces whenever we don't feel like being intimate. Arab women who are forced by tradition or Islamic law to walk around veiled sometimes describe their feelings as ambivalent: on the one hand they feel patronized by men, on the other protected from them.

Women who wear the full-length body cloak called *burqa*, either by choice or because they are compelled to do so, sometimes describe it as "liberating", at least according to a 2010 article in the German news magazine *Die Zeit*.

If people don't have a place to withdrawal, they tend to create one, like children building "caves" under their beds or in the closet. One reason many give for moving to big cities is that there, they are anonymous. This, they feel, is a welcome change from the openness and lack of privacy in the small towns or villages they come from. Truth be told, city dwellers have no more privacy than anyone else, since people there live door on door, but as an individual the feeling is that walking down big city streets, surrounded by strangers, they are just a face in the crowd.

Digital anonymity serves a similar purpose when we are online. This is important in an age of transparency and acceleration, and we should all be very worried about politicians calling for what effectively is the online equivalent of laws banning demonstrators to wear masks. Following the cold-blooded murder of 85 young people by Anders Breivick on the Norwegian island of Utøya in 2011, many politicians in Europe renewed their demands that operating anonymously in Web forums and posting comments under a false name should be prohibited by law and punished. The idea being that Internet users should "put their cards on the table".

More and more, operators of online news sites as well as bloggers are seen to jump on this bandwagon voluntarily. In 2011, Facebook started requiring users to use their real names when signing up. Critics pounced on this, reminding the company that people in places like China or Iran who post under their true names face prosecution by the authorities and even in extreme cases death sentences. The well-known Chinese human rights activist and blogger who writes under the name Michael Anti found his Facebook account canceled because he refused to register under his real name, which is Zhao Zing, for fear of being stuck in jail. Besides, he argued, nobody knew him by his legally sanctioned name. Asked for a statement,

Facebook's press department replied that "real names are part of our culture at Facebook and lead to more responsible use and therefore to more security and trust."

It can also lead to financial ruin, as Kaliya Hamlin, an analyst and expert of digital identity & access management (IAM) who lives in San Francisco and runs a company called Personal Data Ecosystems found out. "Identity Woman", as she is known throughout the industry from the name she adopted for her blog, wanted to register on Facebooks arch rival Google+ and was refused. "People only know who the hell Kaliya is, they only know the Identity Women", she complained. "It even says so on my business cards."

Luckily, Google gave in; in July 2014 the company officially rescinded its real name policy. "We hope that today's change is a step toward making Google+ the welcoming and inclusive place that we want it to be", Google said in a statement.

Agents, avatars and anonymity

Our traditional way of thinking about privacy and intimacy no longer serve any meaningful purpose in a digitally networked world, so what's the alternative? What could replace it? There are several likely candidates. One is the concept of "digital discretion" we have talked about at length in this chapter. Others are anonymity and pseudonymity.

Anyone who wants to be treated with discretion in the digital world, we believe, should be allowed to choose their own degree of exposure and, if they so desire, send a substitute instead; someone equipped and empowered to act as their representative dealing with others in the digital realm.

The word "avatar" has been around long before the 2009 film of that name by James Cameron. The word goes back to ancient India and is used to describe the reincarnate Hindu god Vishnu who is believed to have existed in many different incarnations, each with its own persona, namely the avatar. In almost all Hindu denominations, Vishnu is either worshipped directly or in the form of his avatars, the most famous of whom are Rama and Krishna. Vishnu is immediately recognizable to the cognoscenti in any of his various forms, including that of a woman – unlike avatars in the technisized Western sense of the word, where avatars do not bear any resemblance to the human owners.

Avatars have become figures of ridicule among Internet users due to the regrettable attempt by a company called Linden Lab in 2003 to popularize their MMORPGs (Massively Multiplayer Online Role Playing Game) called *Second Life* as an alternate online reality in which we all don an avatar of our own creation and essentially do things we do in the "real" world, only better. For instance our avatars could fly through the air which many of us would love to do in the here and now, but can't. Large corporations jumped on the bandwagon and spend millions of dollars, like IBM, creating a "virtual company headquarters" and in the process making some early adherents outrageously rich: Payment was in virtual "Linden

Dollars", but these needed to be purchased with real, government backed cash.

As it turned out, most people are happy with their own real lives, thank you, and so they didn't need a second one. News reports soon began to describe Second Life as an "avatar graveyard" because of the many abandoned digital deputies left behind by their owners.

For some reason, many people (especially if the work in marketing) think that an avatar has to be a cartoon figure. Witness the "customer avatars" that became popular on some retail websites a few years ago. Mostly these were made to look like female figures from some kid's comic book, and when they greeted you with "How may I help you", they tended less to resemble an avatar then some kind of silly joke.

A real avatar is something I use to represent me on the Internet, such as the photo on my Facebook page. One of the authors of this book chose, a while back, to represent himself as a small green tree frog, and most people accepted that, albeit often only after leaving behind a few choice comments. Fans of the multimedia game "World of Warcraft" are used to being represented to other players as some kind of fantasy creature, depending on one's status and previous level of achievement. Friends (and we use the word loosely here) know perfectly well whom they're playing with or against, as the case might be.

The right to remain anonymous

At the risk of offending those upright pillars of society who demand that onliners be forbidden to wear the mask, we feel that anonymity and pseudonymity are important attributes of our developing online culture and should not be put at risk out of some populist sense of "decent" behavior. In fact, we agree more with people like Frank Rieger, a so-called "hackdivist" and member of the *Chaos Computer Club* in Hamburg, Germany, who demands that anonymity be elevated to the status of a basic human right. Like the right to remain silent, the right to remain anonymous, he believes, is an essential way to protect human dignity, not to mention security of person and others set forth in the Universal Declaration of Human Rights issued by the United Nations.

The German High Court judges added another basic right, incidentally, when they declared as far back as 2008 that the legal protection of human dignity pertains to the contents of our hard disk drives, as well. In effect, the court said, the computer is a outsourced part of our brain and deserves the highest degree of safeguarding against unreasonable search and seizure.

Thus, the high court effectively put a stop to police arbitrary carrying away disk drives and routinely copying all the data so as to fish for possibly incriminating content. There are things that we all store on our hard disks that are so personal that allowing them to fall into the hands of strangers, for whatever reason, is demeaning. Police need to follow the same procedures as for entering and searching through a suspects house, and any information found that is not pertinent to the case being investigated has to be erased immediately.

As much as the high court's decision deserves our applause for their "technically informed decision" (as the newspaper *Sueddeutsche Zeitung* commented later), there remains some confusion as to where the boundaries lie between our PCs at home or in the office and the World Wide Web and other parts of the Internet. As John Gage, one of the founders of Sun Microsystems famously wrote in 1984, "The network is the computer".

Does this mean that the protection against search and seizure extends to the Web as well? And what about my smartphone of iPad: Is that protected too under the "Computer Right"? What about my set top box and my PlayStation? And the Internet icebox or my bathroom scales, if they, too, are connected to the Internet?

As farsighted as they seemed to by, the German judges have left a lot of questions that still need to be answered before we can hope to achieve our final goal of Digital Enlightenment.

Chapter 8

Information Wants To Be Free!

8 Information Wants To Be Free

...but not necessarily free of charge

It was the mantra, the rallying cry of an entire generation of early Internet users: "Information wants to be free". First formulated back in 1984 by Steven Levy in his book „Hackers: Heroes of the Computer Revolution[12]", the sentence was always open to different interpretations. For some, it meant "free" as in "liberty", a freedom to say and do whatever you want (at least as long as your freedom doesn't infringe on someone else's). But for others, "free" had another meaning completely, namely "free of charge".

In the early days of the World Wide Web, especially, "paid content" was a dirty word, and Internet activists argued for ages about whether "commercialism" should even be allowed online.

Artists, writers and musicians, these starry-eyed enthusiasts claimed, could and should present their collective efforts online for free. To make a living, they would use the Web as a place to advertise: musicians would give away songs for free in order to get people to attend their live concerts; writers would collect a huge following of people who would rush to bookstores (or to Amazon, which was still struggling to get started) and buy their deathless prose. And the best thing was: They wouldn't need a publisher be-

---

[12] Hackers: Heroes of the Computer Revolution - 25th Anniversary Edition Paperback (O'Reilly) 2010

cause they could self-publish cheaply and easily and thus cut out the middlemen who were skimming off all the profits anyway. Whoever heard of a rich author (unless your last name happened to be Grisham, King or Rowling.

Yes, those were heady days, but by now, 15 or 20 years later, we have all sort of calmed down and realized that money makes the online world go round, too.

Black holes in cyberspace

In 2012, there was an excited debate on both sides of the Atlantic about two bills pending in the U.S. congress, SOPA and PIPA. The Stop Online Piracy Act (SOPA) was a United States bill introduced by U.S. Representative Lamar S. Smith (R-TX) to expand the ability of U.S. law enforcement to combat online copyright infringement and online trafficking in counterfeit goods. The PROTECT IP Act (Preventing Real Online Threats to Economic Creativity and Theft of Intellectual Property Act, or PIPA) was a proposed law with the stated goal of giving the US government and copyright holders additional tools to curb access to "rogue websites dedicated to the sale of infringing or counterfeit goods", especially those registered outside the U.S.

Both were seen in Europe and other parts of the world as a particularly dastardly attempt to give extend U.S. laws beyond the borders of the United States and in effect give lawmakers in America the power to regulate intellectual property rights around the world.

Protesters took to the streets in London and Berlin. And even U.S. tech companies such as Wikipedia, Google, and an estimated 7,000 other smaller websites joined the fun, shutting down their websites for days to document their opposition to "U.S. online imperialism".

These "black holes in cyberspace", as at least one newspaper commentator called them, proved big enough to trap lawmakers in Washington who were forced to back down. In January of 2012, the then Senate Majority Leader Harry Reid announced that a vote on PIPA would be postponed until issues raised about the bill were resolved. A few days later in the House Judiciary Committee Lamar Smith postponed plans to draft the bill, stating that "The Committee will postpone consideration of the legislation until there is wider agreement on a solution." Neither has been heard of since.

But let us pause for a moment to think the concept of "free content" through. It implies that commercial enterprises (for instance book publishers, music labels, and movie studios) are supposed to find other ways to monetize content by providing additional services instead of simply charging users every time they download something. The same goes for information created by the users themselves. Actually, Facebook already does exactly that by harvesting user information from their postings and peddling it in an "enriched" format as profiles of user preferences and habits to paying clients who wish to use this information for advertising purposes. These companies are seeking to make a profit, of course, which bothers many members of Facebook and other sites who feel that they are being spied on and degraded to the role of "glass customers".

So what to do? Should content be free or not? This is turning into the great dilemma of the Internet Age: Everyone wants great content, but nobody wants to pay for it. But as soon as enterprises take us by our word and start exploring alternate ways of earning a living we complain, too.

There are, of course, other business models that we could probably all agree to if we put our minds to it. Take the piano player's hat which guests throw money into if they like what he's been playing. Content owners, this theory implies, should copy street musicians, some of whom make good money, whilst those who just make noise are forced to take up some other profession. In fact, this business model has been practiced for years now by software developers and other artists on the Internet, but let's be real here: When was the last time you contributed voluntarily when you downloaded a "freeware" program? And all the hype about "crowdsourcing" and "crowdfunding" can't completely disguise this fact which has more to do with human nature than with anything else.

Jaron Lanier, whom we have discussed in a previous chapter and whose deadlocks seem to indicate a freewheeling attitude towards all things digital, if not marking him as a downright later-day hippie, apparently has underwent a sort of Road-to-Damascus experience. Once a pioneer of so-called Virtual Reality" and an outspoken proponent of Digital Transformation, he now proclaims a neo-capitalist doctrine, blaming collectivism, Crowdsourcing and the Open Source movement on stifling commercial activity and contributing to the decline of the middle-class America.

"To my friends in the 'open' Internet movement, I have to ask: What did you think would happen?", he writes in "Who Owns the Future"[13]. Web businesses exploit a peasant class, users of social media may not realize how entrapped they are, and a thriving middle class is essential to keeping the Internet sustainable: These and other apocalyptical predictions place him firmly in the class of cultural pessimists through the ages. While certainly entertaining, these views do not help us address the real problems facing Digital Enlightenment; quite the opposite.

The new sense of justice

Lanier and others taking part in this discussion suffer from a kind of historic blindness which they share with many politicians and policy makers: people whom we would prefer to hope are aware of their role in making history happen. One of the best examples of this is the debate about copyright and intellectual property. The way many people bandy these terms about one could be excused from thinking that ideas are the same thing as material goods such as homes, cars or a backyard barbecue grills.

---

[13] Jaron Lanier, Who Owns the Future (Simon & Schuster) 2014

They aren't, of course. Property by common legal agreement implies judicial rights to moveable and immoveable goods. The interesting question here is: Do ideas – say, for instance, a song, a poem, a computer program which consists of bits and bytes – fall into the category of "goods" at all, and more importantly in an age where these products of human creativity can be reproduced instantly and indiscriminately, can we really claim power over them in any legal sense of the word? Remember: In the digital age there is no longer a difference between original and copy. The only thing anyone can even hope to control is the copying process itself.

This brings us to the legal concept of "copyright" which, it turns out, is a relatively new one. In 1511, Albrecht Duerer, the famous Renaissance artist and engraver, procured the first legal protection against copy pirates from Maximilian I, the Holy Roman Emperor, for his woodcut "The Great Passion". The term "copy right" first appears in English law with the so-called "Statues of Anne" published in 1557, just a century after the invention of the printing press by Johannes Guttenberg in Mainz. However, it had nothing to do with intellectual property: It simply protected the rights of printers (more specifically: members of the English "Stationers' Company") to publish works to which they could prove (or at least somehow convincingly claim) some kind of exclusive reproduction rights. The statute doesn't even mention the authors, who were paid (if at all) for providing a manuscript.

Copyright, it turns out, isn't there to protect artists at all: it protects publishers!

A recipe for pirating

Shakespeare was one of the biggest victims of piracy, as anyone knows who has seen Roland Emmerich's 2011 film "Anonymous" in which he presents Edward de Vere, 17th Earl of Oxford, an Elizabethan courtier, playwright, poet and patron of the arts, as the true author of William Shakespeare's plays; a theory popular among a certain wacky brand of literary scholars.

Anyone who thinks digitalization is the cause of pirating has got his history wrong. Printers and publishers for centuries have ignored the rights of artists and ruthlessly helped themselves to all the "intellectual property" they could lay hands on. Printers in Shakespeare's day hired hack writers to sit in theaters and make shorthand transcriptions of popular plays such as *Romeo and Juliet* or *Macbeth* which were they published in cheap "quarto" editions. The playwright, be his name Shakespeare or Oxford, never saw a penny.

Modern copyright pirates sit in movie theaters, equipped with small, ultra-high definition handycams, or they smuggle tiny digital audio devices into rock concerts and come away with recordings that rival those made in the largest studios and which can be copied onto DVDs in minutes. The only risk they run is being ejected by the bouncers, who in turn will often turn a blind eye for a small consideration. Or maybe the pirate is actually in league with the owner of the movie theater or concert hall...

Shakespeare may well have been ahead of the times. None of his plays were ever printed officially during his lifetime; the "First Folio" edition appeared long after his death. Instead, he lived quite comfortably it seems from his box office proceedings. The written texts of his plays were, for him, just the raw material from which he created unforgettable theater moments. Which brings us back to Roland Emmerich and his films: Who cares about the film's script? We want to see some action here! Crashing skyscrapers! Gigantic space ships falling from the skies! Emmerich is a master of this genre. As for English literary history – maybe not so much…

Art without copyright – copyright is not an art

Listening to the high-pitched arguments about digital piracy one could almost assume that copyright or commercial legal protection of intellectual property are almost a law of nature; something we always enjoyed and that's been around forever. Not so: Most apologists of copyrights forget that it has only existed for a relatively short time, and that it arose from a certain historical context. There may have been a time and a place for it, but no longer!

John Naughton, professor of the public understanding of technology at the Open University, believes that "Intellectual property is, in fact, just a temporary monopoly granted by governments to authors so they may benefit from their creativity. The monopoly is temporary because it is felt that society benefits from the free circulation (i.e. unrestricted copying) of ideas, and the period of copyright protection represents an attempt to strike a balance between the needs of society and those of authors."

There are those among artists of all genres who believe that the moment the notion of copyright lost its innocence, so to speak, was when we all started giving up our rights to intermediaries like publishers and music studios. These for-profit organizations syphon off almost all the profits, leaving the "owner" of the intellectual property a pittance, if that. And they get away with it despite of the fact that they contribute relatively little to the process besides marketing and distribution.

But maybe artists don't need monopolies after all. An increasing number of them are cutting out the middlemen and going directly to their audiences. This is becoming a much more attractive business proposal in the eyes of musicians such as singer Courtney Love, the widow of Kurt Cobain, who founded the rock group "The Holes". She argue that artists should go back to the business model of accepting tips from listeners, just like street musicians do all over the world. In the age of the Internet which gives artists instant access to large numbers of potential fans, this actually sounds realistic and feasible.

In China, where unchecked digital piracy is the norm, most artists have given up on the idea of making money from CD sales. Instead, for them, recordings serve as cheap form of advertising aimed at roping audiences in to attend their live concerts which pay them enough to live on.

That, by the way, is just how most musicians see radio: they don't earn any money if their songs are aired except the paltry sums collecting societies in some countries bring in. But they do profit immensely by exposure on the airwaves.

Even large acts like Madonna or the Rolling Stones share this economic model, albeit to a lesser extent. The Stones earn more than your average local act through recordings, but by far the largest part of their wealth is generated through their world-wide concert tours which draw giant audiences in football stadiums or open-air arenas. Their content – the songs they sing – are essentially reproducible anytime, anywhere by anyone with a stereo system or an iPod. What makes all the difference is the experience of witnessing a live performance, even if every second on stage is scripted down to the tiniest gesture and facial expression. Concerts, opera and theater performances: In virtually every area of artistic endeavor it's the emotions that count, the feeling of "being there" and taking part in a larger group experience; in short, participation!

One possible exception is literature, where it is harder to create a live group experience. But increasingly, events like "poetry slams" and literary readings on YouTube are helping belle-lettrists to catch up.

Is intellectual property theft?

In an essay for the *Guardian* entitled *Intellectual property is theft. Ideas are for sharing* published in 2003, Naughton writes: „We have to remind legislators that intellectual property rights are a socially-conferred privilege rather than an inalienable right, that copying is not always evil (and in some cases is actually socially beneficial) and that there is a huge difference between wholesale 'piracy' – the mass-production and sale of illegal copies of protected works – and the file-sharing that most internet users go in for."

Sharing is actually as old as art itself, as is borrowing from others. Witness Brahms and his *Variations on a theme by Haydn*, or Plato's *Dialogues* which builds on ideas from Socrates. Only in modern times has plagiarism become a plague, something to be pounced upon and denounced.

The idea of "protecting" intellectual property and outlawing copying is of recent origin, and its proponents do not sufficiently understand or take into account the historic background. Since nobody remembers what the problem was we are trying to solve here, laws are passed criminalizing an entire generation of young people who are growing up in a world where "sharing" is a virtue and a value in itself. We need to change these laws to take into account where the digital society is heading.

The same goes for patent laws. Patents are nothing but a technical variety of copyrights. If proof were needed we must only point to the abstruse and apparently never-ending "patent wars" between companies like Apple and Samsung and their ilk. Apple actually has patents that cover "esthetic appearance" of smartphones; things like design color and form. Their lawyers have actually gone to court to prohibit competitors from making phones with "four evenly rounded corners", "a flat, transparent surface" and a display that "is centered beneath a transparent surface", as well as a product that, "when turned on, shows colored icons on its display."

This sounds crazy, but there is method behind it. The object is to stop others from making something that is better than what Apple offers today. Which is another way of describing the end of progress in technology and design.

Contrast this with a statement issued by WIPO, the *World Intellectual Property Organization* that patents "simultaneously fosters innovation and remains consistent with fair market rules." This is humbug: patents stifle innovation by their very nature! This is especially true in computer technology. As Richard Allen Posner, an American legal theorist, economist, and former judge on the United States Court of Appeals in Chicago, wrote in his blog:

"Nowadays most software innovation is incremental, created by teams of software engineers at modest cost, and also ephemeral—most software inventions are quickly superseded. Software innovation tends to be piecemeal—not entire devices, but components, so that a software device (a cellphone, a tablet, a laptop, etc.) may have tens of thousands, even hundreds of thousands, of separate components (bits of software code or bits of hardware), each one arguably patentable. The result is huge patent thickets, creating rich opportunities for trying to hamstring competitors by suing for infringement—and also for infringing, and then challenging the validity of the patent when the patentee sues you."

Patent trolls are a particularly virulent form of quasi-monopolies, defined by Wikipedia as "a person or company who enforces patent rights against accused infringers in an attempt to collect licensing fees, but does not manufacture products or supply services based upon the patents in question, thus engaging in economic rent-seeking." Like the lilies of the field, they toil not, neither do they spin.

Ironically, one of the fiercest patent litigators, Apple, is itself the target of patent trolling, or so the company asserted in two friend-of-the-court briefs for cases pending in the U.S. Supreme Court submitted in 2014. In them, they claim to had to face nearly 100 lawsuits in the preceding three years.

In Europe, discussion is underway about way to reform copyright and patent laws to bring them in line with the cultural changes going on as digitalization and networking increasingly reshape our collective sense of justice in society. In the journal "Le Monde diplomatique[14]", the Dutch art teacher Jost Smiers wrote: „The concept of copyright protection, once a consistent and convincing idea, has become an instrument designed to give a handful of mega-corporations control over tour common intellectual property." 90 percent of money earned through sales of intellectual property of all kind, he maintains, land in the pockets of publishers and studios; writers and performers only get about 10 percent. Which just

---

[14] Le Monde diplomatique , No. 6549

goes to show how phony the claims are that copyrights are there to protect the poor artists.

The effect of these monopolistic control structures is catastrophic, Smiers says, since they guarantee that only art and entertainment owned by large corporations has a chance at reaching the general public. "They are only interested in creating a handful of mega acts and superstars in which they invest huge sums and who earn them fortunes in merchandizing and marketing. Because they risk losing their investment they pursue aggressive worldwide marketing strategies aimed at displacing any alternative, any artist they don't own, from the public consciousness." The result, of course, is cultural poverty.

Old content, new context

In art, copying someone else's work is called "plagiarism", and it has a bad name. Which is actually surprising, since it's been done ever since art was invented. Cavemen presumably copied the wall drawings of their fellows. "Through all of the history of literature and of the arts in general, works of art are for a large part repetitions of tradition", Wikipedia notes. "Immature poets imitate; mature poets steal", T.S. Eliot is quoted as saying. With the rise of "cut & paste", copying has become so widespread that teachers and university professors now use "plagiarism checkers", special software designed to catch out students who cheat on their term papers by helping themselves to someone else's online texts.

But is copying really cheating? Or is it in fact a form of art?

When Axolotl Roadkill, the debut novel by Helene Hegemanns, was published in 2010, a huge literary scandal erupted when the young German writer freely admitted that she had copied entire passages written by an anonymous blogger who went under the name "Arien". Instead of being embarrassed, Ms. Hegemann dismissed critics by claiming to have "re-contextualized" the text. In her eyes (and in the eyes of many of her fans and followers) this is a legitimate artistic technique, akin to remixing, sampling and cover versions popular among musicians, especially in the realm of New Age and Electronic Music pioneered by such groups as Kraftwerk, Yellow Magic Orchestra, and the Frenchman Jean Michel Jarre.

In the 80ies, the Hiphop artist whose chosen name was Grandmaster Flash was taken to court for having "sampled" some the baseline from an album by the Sugar Hill house band for his hit song Whitelines. A German court rejected the case, stating that Flash had created a new work of art; the sequence he used only served as the raw material for a new "original", just like the clay a potter uses to form a vase. "My job is to create a track", Flash stated in an interview with Rapneck Ossi and Ziggic Moondust for in their book HipHop (attentive readers will have noticed the subtle reference to pseudonymity discussed in chapter 7). "Scratching and all that stuff is how I fill the track. Rap takes something that's already out there and rebuilds it, gives it a new structure and adds the rap on top."

Just how hard it is to prove plagiarism was noted by the French philosopher Jacques Soulillou in his book L'auteur mode d'emploi (roughly: "Instructions for the Author"). "In art and literature it is difficult to demonstrate that B has borrowed from A without quoting the source because it would first be necessary to find out whether A has lifted it from someone else entirely. Plagiarism requires that the process ends with B copying from A." If the chain extends further back, he implies, then it isn't plagiarism at all.

We believe it's possible that the creative process always involves some kind of plagiarism, or borrowing from others. The term "recontextualizing" describes this very well. An artist can only work with things he or she has experienced or learned from others. Originality therefore is a myth: No one is an original thinker! Einstein constructed his famous Theory of Relativity with the help of knowledge acquired by generations of mathematicians and physicists before him. Or, as Isaac Newton famously said: "If I have seen further it is by standing on the shoulders of giants."

Or to put it bluntly: A good copy job is itself a work of art. Only poor copies deserve to be stigmatized as plagiarism. This is much more value statement than a judgment, much less a legal ruling. In future, we believe, no one will call a copyist a crook. Instead, we will think: "This guy is a plagiarist, how boring!" And since nobody will buy or use such a product, the punishment for plagiarism will be to be ignored.

Good plagiarists, on the other hand, will be in great demand, and people will be willing to pay for their productions. When Emerson, Lake & Palmer released a rock version of Aaron Copeland's Fanfare for the Common Man, they sold millions of copies of what had in effect become a new original work of art.

In his book Mashup, Dirk von Gehlen who we mentioned above explores the idea of differentiating between plagiarism (replicating with intent to deceive) and copying (with intent to disseminate). Sounds like an interesting idea. Maybe we should borrow it...

Information without context

Journalists, of course, belong to a profession that has always used copying as a tool of their trade. "A good ideas doesn't care who had it", is old hack slogan.

Naturally, our journalist friends don't like to hear this. Instead, they launch self-righteous attacks on any scientist or politician stupid or unlucky enough to be caught copying, which isn't that hard given the advances in software that scours the Internet in search of matching texts.

The irony is that journalists themselves are the biggest plagiators of all. One could even say that's part of their business model. Reporters are the quintessential hunters and gatherers of the news trade. They are constantly on the lookout for information that they don't make up themselves (exceptions prove the rule) but instead "research", or dig out. Often the originators of this information would prefer they didn't, and then we call it "investigative journalism".

Besides reporting the facts, journalists also write comments which, at least among professional journalists, are supposed to be strictly separated from one another. If not, there is the danger that reporters will grind their own axes instead of focusing on finding an objective version of the truth.

While the separation of reporting and commentary is a traditional value in the Anglo-Saxon media, journalists in other countries have often followed their own or their employer's agendas. In central Europe, newspapers in the 19th and early 20th century were generally published by political parties or groups with strong cultural or party-political roots, if not by governments themselves. The "journalists" they hired were usually drawn from the ranks of their followers, and they tended report only those facts that fit their world view. These carefully vetted "facts" were then placed in the desired context, so in effect they were being more or less "recontextualized"; a term that resonates today not only in the blogosphere, but in many professional forms of media as well. Take *Fox News* as a good example today.

Historically, journalists are the direct descendants of the storytellers who entertained the tribe gathered around the campfire by reciting the legends and sagas brought down to them for their predecessors. They rarely invented the stories themselves, they just carried on an old oral tradition. By the time they were finally written down, the stories were already firmly established in tribal memory, like the legends of the Bible or the Norse sagas.

The tale of the Great Flood probably originated, as we now assume, among the ancient Sumerians long before Noah. The Gilgamesh Epos in ancient Babylon described just such a deluge sometime around 2,000 B.C., and it crops up in the Puluga Saga popular with indigenous inhabitants of the Andaman Islands and in the Popol Vuh, a series of mytho-historical narratives told for centuries by native habitants of modern-day Guatemala.

The journalistic craft is exactly that: a skill that used to require a certain deftness with handling glue. Within living memory (or at least our own lifetimes) press releases were cut apart with scissors and reassembled together with added bits and pieces often handwritten by the journalist into complete new articles which were then sent off to the typesetter.

Nowadays, the same thing is accomplished digitally with the help of a technique called, appropriately, "cut&paste", and it is generally frowned upon by professional journalists. Nevertheless, it still persists, especially since newspaper publishers have started cutting down on paid reports and correspondents in some of the remoter corners of the world. Much cheaper to simply "recontextzualize" bits and pieces from agency reports or, increasingly, tweets reporters on the ground send back via Twitter. Rearranged and embellished with comments and additions by the editors, it look just like somebody filed a full report from the front.

But why, one asks, do readers need someone to assemble their news for them, anyway? Why not go to the original sources which are all available online anyway?

If Google is able to assemble a choice of news items "on the Fly", even making sure that they are all about stuff I'm interested in and like to read, using a computer algorithm to do the report's job, what's the real difference?

In the early 90ies, scientists at MIT Media Lab in Boston developed what they called the "personal newspaper" which drew on publically available news sources which were assembled into a digitally formatted "newspaper" which could be printed out and read just like a real newspaper. The big problem was getting people to buy printers big enough to produce a something the size of a real daily like the New York Times or the Boston Sun. For some reason, these scientists still believed, people will prefer to hold something in their hands that reminded them of the good old daily rag.

Today, of course, more and more of us have become used to reading our "paper" on a computer screen or even on our iPads or smartphones – something even scientists at MIT could imagine just a decade or two ago.

Bloggers –amateurs take over the newsbeat

More and more often, the tidbits of new many of us assemble into their own personal patchworks of information no longer come from professional journalists, but from rank amateurs. Bloggers invest lots of time and care into writing online diaries that are sometimes followed by thousands or even millions. And we, the pros, ask ourselves why?

Sometimes, these amateurs turn pro themselves and start making real money. The *Huffington Post* was reportedly acquired by AOL in 2011 for 315 million. And even smaller blogs now sometimes sport paid advertising banners no longer is necessarily an unprofitable activity by no means.

The danger, of course, lies in the fact that advertisers and other financial backers can influence the way a blogger slants his facts without it being apparent to the readers. Many bloggers agonize about this, arguing the pros and cons of accepting paid ads in online discussion forums and "blogger conventions" regularly hosted in many countries.

Just like professional journalists, bloggers risk ruining their reputations by appearing to be in the pockets of vested interests. Some therefore refuse the temptation; others accept but just don't talk about it. A study published in 2013 by Edelman, the world's largest PR company based in New York, claims that only 26 percent of respondents in developed countries believe in the credibility of news they find on the Social Web, as opposed to 51 for traditional media. Interestingly, almost the same number of people (47 percent) believe what they are told (or pointed to) by search engines like Google. On the other hand, 47 percent of young people aged 18 to 29 appear to use social media websites as their primary source of news and other information.

On the other hand, more than 40 percent of consumers say that information found via social media affects the way they deal with their health, according to *getreferralmd.com*, an online health portal. "Liking, following, linking, tagging, stumbling: social media is changing the nature of health-related interactions", writes Karla Anderson, Principal at PwC, a consultancy. If people trust bloggers with their lives, something pretty drastic must have happened.

There are increasing attempt to regulate or at least police the Social Web to weed out those who abuse the trust users place in them. In America, the National Advertising Review Council (NARC) founded a working group called „Electronic Retailing Self-Regulation Program" (ERSP) which is tasked with uncovering unhealthy links between bloggers and advertisers.

NARC was reacting to reports that a blog site named *WeKnowDiets.org*, which was heavily promoting a product called "MiracleBurn made by Urban Nutrition Inc., was in fact a front for the company which was promising bloggers 20 for every endorsement. The ERSP found that the blog was giving "the impression that there was no affiliation between the review and the company" and called on the manufacturer to make full disclosures to comply with accepted rules by the Federal Trade Commission (FTC) so that consumers would be immediately aware of the connection upon visiting the website. A recent visit to the MiracleBurn homepage showed that the page has been effectively shut down.

The ERSP also singled out a company called Herbal Group which was popularizing its product "Prostalex

Plus", a food additive, on a blog site called Prostate Health Blog which created the impression that this was actually a bone fide drug. The watchdogs managed to force the company to shut down the blog for a while, but it soon opened again with a small link added which led to Herbal Group's own corporate website.

Of course, there isn't actually a law that forbids companies from financing bloggers, but that could change. The FTC has threatened repeatedly to crack down on dishonest online advertising in general and bloggers who accept unacknowledged financial sponsorship in particular. Advertisers, understandably, disagree. "The FTC doesn't understand the medium", Linda A. Goldstein of the attorney firm Mannat, Phelps & Phelps which represents NARC was quoted by the *New York Times.*

The twilight of journalism

People living in the age of networking and digital acceleration have more information at their fingertips than at any other time in history; too much information, in fact, as cultural pessimists keep reminding us. The image they conjure up is of a gigantic tsunami of bits and bytes that threatens to overwhelm us. Those that warn the loudest are usually those who used to be in charge of doling out information in easily digestible bite-sized portions. They call this "quality journalism" and talk of "journalistic care" with which they (or perhaps more accurately: the owners and managers of the publishing houses or broadcasting studio they work for) have deemed appropriate for the general public. Like the Delphic oracle of old they reserve the right to interpret the signs of the times, extracting the true indelible truth from the confusing mass of facts and fictions that constitute the stream of information, both analog and digital, with which we are all surrounded. This they proceed to proclaim in the editorials and commentaries so that we simpler folk can better understand the world.

For this they were once well paid, but alas no more. Print media everywhere are in chronic decay, with circulation plummeting almost as fast as ad revenues. This is bad news especially for the hunters and gatherers: publishers for years have been reducing the budgets for freelancers to the point where some actually argue that being allowed to submit an article for free is good for the author because it gives him "exposure". Journalists, especially those no longer able to hold down a steady job, are rapidly joining the ranks of the "precariat", as those on the margins of subsistence are now being called.

This is not the fault of the journalists, but of their bosses who completely failed to see what was coming and still remain for the most part ignorant of the way the news business in an online world works. For years they persisted in giving away the content for free under the term "recycling" content" which, they assumed was worthless unless printed on paper.

The result is a steady decline into redundancy. Prof Wolfgang Henseler of the University of Passau has charted the development of newspaper circulation in Germany and found that by the year 2034 at the latest the last large daily newspaper in Germany will have been driven out of business. Tabloids will probably predecease them: The mass-circulation daily BILD has seen its print run fall from 3.96 million in 2002 to 1.44 million a decade later.

This doesn't just have consequences for the business model of publishers; it is also a blow to their self-esteem. It means that the media czars are losing control of the message --- and with it their *raison d'etre*.

The whole *endzeit* bellyaching about the dwindling power of the "fourth estate" really only masks the inability of publishers and broadcasters to deal with their own marginalization in the eyes of readers and viewers. Who cares what the New York Times things is all the news that's fit to print: That is for the readers to decide, and they don't even have to print it. And young viewers have long ago switched the channel: If they watch TV new at all, then Jon Stewart's *Daily Show* which is simply a parody of the ego-driven displays of vanity like the *CBS Evening News* or *Nightline*.

Chapter 9

The Great Earthquake of New York

The Great Earthquake of New York

The social upheaval described here is not something dimly perceived on the horizon: it is already well underway. We all should be excited to find out what it will lead to and who will be the most affected. Not that there will be any formal oral or written exam to prove whether we have passed the test of Digital Enlightenment. Instead, the rite of passage will look more like an online game or a YouTube video. At any rate it will involve interaction between growing numbers of people who are beginning to share a number of key formative experiences, both cultural and social, that enable us to gauge our digital maturity and practice new habits and modes of thought.

History is full of such key experiences. Andreas Zielcke, the former culture editor at *Sueddeutsche Zeitung*, a leading German daily newspaper, wrote an article in 2013 in which he drew a parallel between the NSA scandal which revealed the extent of random American government spying on individuals around the world and the earthquake that struck the Portuguese capital of Lisbon in 1755. At the time, Lisbon was one of the most beautiful cities of Europe, and when it was destroyed the event shocked people all over Western civilization, causing them to question their faith in divine providence. "The catastrophe robbed many of their confidence in God's mercy in ways that resemble the terror caused by the collapse of the Twin Towers on September 11, 2001"", Zielcke wrote, and he quoted Goethe who described his own reaction to the Lisbon earthquake like this:

> *"Perhaps the demon of terror has never before sent its harsh shiver over the world more swiftly ... God has proven himself anything but benevolent."*

At the time, events in Lisbon had broad philosophical ramifications, as well. No longer was it possible to base your beliefs on the assumption that this, as Gottfried Wilhelm Leibnitz put it in his *Théodicée* is truly the best of all possible worlds; a notion that Voltaire famously ridiculed in his satire *Candide*. Thus, the Earthquake of Lisbon serves as a handy way to date the beginning of the 18th century European Enlightenment of which Voltaire was a founding member.

The fear of freedom

Historians of the 21st century may well date the beginning of Digital Enlightenment from the collapse of the Twin Towers in Manhattan which have had a truly earth-shaking effect, upsetting Western civilizations belief in one of our most basic human freedoms, the right to free communication. Just so, the Great Earthquake of Lisbon in 1755 shook society's belief in divine providence, thus contributing to the subsequent era of European Enlightenment.

Just how sharply this "Earthquake of Manhattan" has changed the United States is evident if you look at the very different reactions on both sides of the Atlantic to "NSAgate", namely the global surveillance disclosures by the ex-NSA contractor Edward Snowdon. While Europeans were shocked, in America even liberal commentators such as Tom Friedman (author of *The World is Flat*) felt compelled to state in a column for the New York Times that he, personally, would be glad to sacrifice his right to privacy if it would keep him from becoming the victim of a second 9/11. Right-wing political analysts like Ralph Manning even went so far as to demand the death penalty for Snowdon's "act of treason".

Behind these scary rantings lies a cunning logic: Since terrorists, too, use the anonymity of the Internet to disguise their devious dealings, spy agencies and law enforcement officials need to know everything that goes on in cyberspace including illegal or quasi-legal snooping on the private conversations of millions of citizens both abroad and at home. As our friend Sebastian von Bomhard put it in a blog post, governments use statistical arguments to justify invasions of privacy: "What would you prefer? Should we read all your mails or would you rather watch your family die in a terrorist attack? A, B or none of the above?"

9/11 has become the most popular way to rationalize a return of the snooper state by turning the fight for human rights on its head: Instead of being ashamed of what they're doing, these digital eavesdroppers use fear as a new angle of attack. Who will protect us from our protectors, the Roman philosopher Juvenal asked two thousand years ago. Has anything changed since?

What these intractable security experts fail to realize is that they are themselves in the process of irreversibly throwing Pandora's Box wide open. By trying to turn back the clock on digital transparency, they are proving that there are no secrets any more. It is one of the great ironies of the digital age that it needed an Edward Snowdon and 9/11 to fully demonstrate this.

Pity the messenger

Unfortunately, this fact isn't generally recognized yet. Instead of heeding the signs of the times, politicians persist in trying to plug the holes in the system by unleashing the brutal power of the state on those who, like Edward Snowdon, have dared to turn the light on their sinister machinations; about as classic case of shooting the messenger as ever was.

It isn't the short-term, knee-jerk reactions by the authorities we worry about. These were to be expected after the world-shaking events of September 11. It is the long-term shift in the balance between protections against terrorist activities on the one hand and the hard won Fourth Amendment right of free citizens to protection against unreasonable searches and seizures which applies as well to the contents of all communications, as has been repeatedly recognized by U.S. courts.

While the U.S. government persists in branding Snowdon a traitor and a criminal, citizens around the world almost universally agree that he is in fact a freedom hero. And even in the United States, a majority believes he is a legitimate whistleblower, as a survey by the Quinnipiac University in Connecticut proved in July 2013.

However, the term "whistleblower" itself is ambivalent, carrying the positive association with "informer" as well as implying the fouling of one's own nest. According to the scientists involved in the Quinnipiac study, the former meaning was understood to apply by most participants.

But in fact things aren't that simple. The answers to the questions posed by the researchers also revealed something that can be described as the "modern security paradox": while most people share the desire for informational self-determination and freedom from illegal snooping, they equally want protection for themselves and their possessions. 54 percent of those surveyed felt that surveillance was necessary "to make America safer", while 53 percent believed that the surveillance programs uncovered by Snowdon and others went too far and constituted "invasion of the privacy of American citizens".

It is interesting to note that the dividing lines between the two opposite values do not follow traditional party orientation. It is simply not true that Republicans and Democrats are widely separated on these issues. The fear of government snoopers run amok is consistent throughout the political spectrum. Who, indeed, will protect us from our protectors?

The legacy of the Twin Towers

If the fallout from the destruction of the World Trade Center in 2001 is truly comparable with that from the Great Earthquake of Lisbon, we need to ask ourselves whether its long-term effects, too, will be as positive as those from 1755. By triggering the European Enlightenment (or at least substantially boosting it), the events in the Portuguese capital were beneficial for most of mankind. In other words: Could the terrible attack on the Twin Towers lead to something similar?

Andreas Zielcke, with whom we exchanged thoughts on this issue via e-mail, is convinced that the answer is yes. He writes: "Unlike 1755 the learning effect today is two-pronged. Not only must we learn from 9/11 how much hostility can be released by fear of modernity, but we must also learn to curb our own overreaction to these fears; a response that tells us much about modern human nature. Apparently we are perfectly willing to trade in fundamental freedoms for an increased sense of security. On the Internet, we exchange them for various kinds of digital service with companies like Google, Microsoft or Apple. In the process, the value of freedom is diluted; Thomas Friedman's position you mention tells it all. Back in 1755 people learned a disheartening lesson about God and the established church, today we are forced to plumb the depths of human depravity. Yes, Snowdon's disclosures have started a debate, but unfortunately the media and politics have so far failed to respond in an enlightened way to the rise of the surveillance state. We should not just hope for, but instead actively demand public self-enlightenment that will bear fruit in the future. We not only need international conventions: above all we need a more highly developed Internet and freedom culture."

In fact, the process of self-enlightenment Zielcke asks for is already well under way, and it is proceeding at true Internet speed --- or at least much faster than the customary process of political opinion building. According to a study released in 2014 by the analysts at Edelman trust in government has plunged to historic lows not only in America, but worldwide. As a result, the trust gap between government and business is now the largest ever. The drop in government trust among informed publics was even more dramatic on a country level, plummeting in the U.S. (16 points to 37 percent), France (17 points to 32 percent) and Hong Kong (18 points to 45 percent). Populist sentiment is evident in the fact that among the general population trust in government is below 50 percent in 22 of the 27 countries surveyed, with strikingly low levels in Western Europe, particularly in Spain (14 percent), Italy (18 percent) and France (20 percent). And while the study does not say so explicitly, it is probably safe to assume that Edward Snowdon and the NSA are major reasons.

Watch the watchers!

If the Internet stands most things on the head, why not surveillance, too? After all, if newspaper readers can become bloggers and couch potatoes turn into

YouTube producers (not to mention consumers becoming product designers, see chapter 6), why shouldn't the empowered citizen be allowed to watch the watchers? And isn't it the moral duty of every responsible citizen to keep a watchful eye on elected governments?

Technology has made huge advances towards enabling all of us to join in doing just that. As early as 1991 George Holliday made history by capturing the brutal and unprovoked beating of Rodney King, a black, by white policemen in Los Angeles, thus earning himself the honorary title of the "first citizen journalist", according to a documentary directed by Roby Massarotto. Today, thanks to YouTube, pictures like those of Mike Brown lying dead In Street at Ferguson, Missouri, go viral in minutes, provoking mass protests all over America.

In Berkley, California, the activist group *CopWatch* has for years been recording instances of police brutality towards innocent victims and publishing the cases online. In France, where tensions between authorities and immigrants in the slums of the "banlieus" go back a long way, a similar project entitled *Copwatch Paris Nord I-D-F* went so far as to post the names and addresses of policeman caught on camera racially abusing detainees on their website, leading a court in Paris to order the website to be shut down. Of course, pictures and videos immediately started cropping up elsewhere on the Web, leading to new police raids and to a kind of digital version of whack-a-mole between activists and authorities.

France, the nation that gave us the Universal Rights of Man, has been a particularly grim example of authoritarian crackdowns on human rightists and online free speech in recent years. Paris, it seems, seeks to emulate China in constructing a Great Firewall to keep dissenters offline. Presumably their ancestors who manned the barricades of the Commune are spinning in their graves.

That citizens are willing to take to the streets to voice concerns about digital censorship and surveillance is actually a rather new development, but it is growing fast. The attempts to pass ACTA, an international anti-counterfeiting trade agreement, first led in January 2012 to protest marches in, of all places, Poland, from where they quickly spread all over Europe. In Munich, more than 10,000 protesters gathered despite sub-zero temperatures to vent their anger at the treaty, the text of which remains top secret in the United States for reasons of "national security".

It is interesting to note that resistance to digital snooping by the state is growing even in the U.S. congress, where Justin Amash, a Republican with libertarian leanings, introduced a bill in the House a few weeks after Edward Snowdon's flight to Moscow sharply limiting the NSA's ability to tap phone calls by American citizens. While the proposed legislation went down with 217 to 205 votes, it did signal that a substantial number of lawmakers are sufficiently worried about the abuse of power by the authorities to start a serious debate on the freedom of communication in the digital age. Hey, it's a start!

The NSA, by the way, defends its snooping program by arguing that the data and metadata collected by them are not being read by actual human beings but by machines instead. In a legal sense, they say, nobody is reading them at all.

This is absurd, as the mathematician David Hilbert demonstrated back in 1900 in a paper he submitted to an international math congress in Paris in which he listed the world's "greatest unsolved mathematical problems", among them one he called the "decision problem". This consists of the question whether it would be possible to devise a systematic mechanical process for determining if a given group of symbols convey meaning or not.

In a fascinating article commissioned by the German newspaper *Frankfurter Allgemeine Zeitung* in 2013, George Dyson, a science historian and author of *Turing's Cathedral*, claimed that the answer is a resounding "no". "In modern computational terms", he wrote, "no matter how much digital horsepower you have at your disposal, there is no systematic way to determine, in advance, what every given string of code is going to do except to let the codes run, and find out." U.S. spy agencies trying in effect to use computer algorithms that can read our minds are doomed to fail because they will fall foul of Hilbert's decision problem.

Dyson concludes from this that "it will never be entirely possible to systematically distinguish truly dangerous ideas from good ones that appear suspicious, without trying them out. Any formal system that is granted (or assumes) the absolute power to protect itself against dangerous ideas will of necessity also be defensive against original and creative thoughts. And, for both human beings individually and for human society collectively, that will be our loss. This is the fatal flaw in the ideal of a security state."

Fear of strange thoughts has always been a distinguishing factor of authoritarian societies. To avoid falling into this trap we need to allow thoughts – and data – to flow freely in order to find out, as Dyson suggests. In this context, the concept of "go with the flow" already mentioned in chapter 2 gains an unexpected political dimension. Openness and transparency, two vital and inalienable values on our way to Digital Enlightenment, therefore come with a price tag attached. They force us not only to accept dissent but even to be willing to live with the risks inherent in total freedom of communication. Openness and transparency, while desirable in themselves, are also sometimes opposites that require a delicate balance if we don't want to end up living in a self-incarcerating society.

Chapter 10

The End of Utopia

The End of Utopia

My interest is in the future because I am going to spend the rest of my life there.

Charles Kettering (1878-1958)

Just a decade and a half ago, in 1998, the chancellor of Germany, Helmut Kohl, was asked by a journalist what he thought about the rise of the "Datenautobahn", the data super highway that was the Internet. He replied: "You'll have to ask the folks at the Ministry for Transportation".

Around this same time, a radio station in Bavaria broadcast a traffic warning concerning a problem the Internet provider Deutsche Telekom was having with their mail servers, which had led to a "data jam on the information highway." Drivers on their way to work presumably were slightly confused.

There are enough funny quotes about the Internet to fill a completely different book, but the point is that politicians often betray a disturbing lack of knowledge about digital technology in general and the Internet in particular. When protests against corruption and nepotism in his government escalated early in the summer of 2013, the Turkish premier Recep Erdoğan reacted by ordering his flunkies to shut down the servers at Twitter, which he termed a "plague". And during the wave of demonstrations following the allegations of fraud during the 2009 presidential elections in Iran, a foreign correspondent's car was stopped and searched by members of the Revolutionary Guard who demanded to know if he was carrying a copy of a forbidden tome called "the Facebook".

We ourselves remember a conversation with a young student from Egypt who, while the protests against the regime of Husni Mubarak were going on in Cairo, was looking in Munich for someone interested in renting out an unused communications satellite which he proposed moving into a geostationary orbit high above the Egyptian capital in order to provide uninterrupted Facebook access to the thousands of young people gathered in Tahir Square. The government, it seems, was periodically shutting down the Internet servers in an attempt to stop the protesters from organizing new demonstrations. Lucky, Mubarak was deposed before the young man could find anyone to do business with, thus presumably saving himself a lot of money.

Politicians more than anyone apparently need some Digital Enlightenment, and that as soon as possible. Like everything else, the Internet is rapidly changing their business model, too. The political decision making process could conceivably soon be taking place in real time, and why not?

On the other hand, attempts to silence the growing digital unrest in places around the globe are becoming more and more frequent. Not that they are very successful, but the question remains: Why do politicians keep trying if they know it won't work anyway?

Politics and policymaking are subject to the same laws of digital acceleration we have encountered elsewhere in this book. Society is changing, and pressure is growing for political reform that will empower the individual and speed up the political process itself. A small but growing number of politicians realize this and are adding to the chorus of voices demanding systematic change. But what is Digital Democracy and what's the difference to good, old-fashioned analog politics as usual?

Politics in real time

The purest form of democracy is one in which every citizen casts his or her vote and the majority wins. In the Internet Age, at least theoretically, everyone in the world could participate in the democratic process which could be vetted and overseen by computers programmed to detect any kind of fraud, including anybody tampering with the computers themselves.

Given that, the question is: Just how far can direct democracy go? The most radical answer of course is all the way! Digital democracy could potentially replace party politics and representative government with a participatory system giving individual citizens a direct and active role in the political decision making process.

In Europe, attempts by the short-lived "Pirate" parties have shown how difficult, not to say impossible it is to combine the concept of direct democracy with existing party structures. As long as they stayed out of the parliaments, the Pirates were an effective force driving change at many levels. Once they entered the mainstream, they imploded in the kind of petty bickering and horse-swapping deals that mark the ruling political castes. Instead of leading the way, the Pirate parties have receded into part of a vague and largely ineffectual "oppositional counterculture" that has been around for years and has, at times, been highly successful in shaping the political agenda.

The Pirate movements serves as a warning about what happens when you try to adapt truly innovative concepts to existing old and declining structures without first thinking about how these structures restrict and stifle innovation. Any attempt at introducing direct democracy into a party-driven political system is a sure recipe for failure.

Much more interesting (and more likely to succeed) are attempts such as those of the online activists going under the name of *Anonymous*. In their book *Anonymous: Pirates informatiques ou altermondialistes numériques?* (Anonymous: information pirates or digital anti-globalization?), authors Frédéric Bardeau and Nicolas Danet describe how this group of self-proclaimed "hacktivists" evolved into a kind of online offshoot of leftist activism. Through their often spectacular actions and initiatives they quickly earned themselves a worldwide reputation and could very well become a blueprint for how political decision making will be shaped in the digital future as well as for how the Internet can and will be used as a platform for political action.

In the past, Anonymous has targeted such diverse opponents as Scientology and Sony, the Mexican drug mafia and pedophile networks, as well as FBI and NSA and of course repressive regimes in places such as Tunisia, Iran or Egypt. During the 1999 World Trade Organization summit in Seattle, the group played a key role in what later became known as the "Battle for Seattle", circulating plans for the shutdown of the city on the Web and coordinating attacks by activists against municipal and federal authorities. The event is recognized today as the start of the anti-globalization movement in the U.S. and elsewhere. Derided by the traditional media as "anarchists" and "criminals", Anonymous has subsequently been hailed by the left as digital freedom fighters.

Bardeau and Danet draw a remarkable historic parallel: "Could Anonymous, which sprang from an open Internet based on the principle of cooperation, just as the European Enlightenment can trace its roots back to the invention of the printing press, be the spearhead of a global revolution?"

So should we abolish political parties and start over again online? Are parties, as George Washington warned, destructive because of their temptation to manifest and retain power? In his 1792 Farewell Address he states that *the common and continual mis-*

chiefs of the spirit of party are sufficient to make it the interest and duty of a wise people to discourage and restrain it.

A somewhat less radical approach calls for creating a system by which voters can participate directly in reaching choices that effect everyone, something that has been practiced for centuries in Switzerland. The Swiss constitution calls for plebiscites in certain substantive issues that are deemed too important for mere politicians to decide.

Perhaps it would be better to think in scenarios instead of calling for the abolishment of political parties altogether. One could be that politicians and parties finally manage to agree on necessary changes to society and the body politic that would enable direct forms of democracy in certain instances --- albeit with the risk of rendering themselves superfluous.

Another could be the model of oppressive "formatted society" like the one described by the German historian Wolfgang Mommsen in 1999, when he painted an gloomy word picture of the future in a book published by the social-democrat Hans Eichel and the cultural functionary Hilmar Hoffmann entitled *End the State – Start the Citizen Society*. Direct participation of citizens in the governing process, he warned, would need to be stamped out completely in such a society since only representative democracy is able to guarantee the necessary stability. This argument, of course, isn't new at all; it echoes, in fact, the model of government through benevolent dictatorship that

Thomas Hobbes suggests in his in 17th century masterpiece, *Leviathan*.

These two scenarios represent opposite extremes, but both they and anything in between are conceivable. Political parties do pay occasional lip service to direct democracy on both sides of the Atlantic, but nothing much ever happens. On the other hand, examples from Scandinavia and Switzerland seem to point in the other direction and give reason for optimism.

The main problem that opponents of greater participation by the citizen are fond of quoting is a practical one. Nobody, they argue, wants to turn out weekend after weekend to get to the polls in order to vote on some petty local issue. Besides, there is the generally accepted idea that we need a stable administration in order to conduct the day to day business of the state. It would be impossible, these people argue, to assemble a government on the fly, kind of like rolling dice.

But the real reason professional politicians and party hacks fear participatory democracy is simple. Barbara Prammer, the president of the upper chamber of the Austrian parliament, recently gave the secret away unwittingly when she said in an interview: "The nature of democracy is compromise. I don't like these 'all or nothing' votes that are typical in a direct democracy. "The occasional referendum, that's okay, she believes. But in the end it's up to parliament to make the decisions.

This of course is direct democracy with support wheels: let the people think that they're in charge, but leave the important stuff up to the patronizing politicians.

In neighboring Switzerland, Ms. Prammer could see a kind of direct democracy that actually works and has been doing so for centuries. One of the peculiarities of the Swiss system is its consensus principle by which all the major parties agree to form a government, dividing up all the portfolios amongst themselves according to a proportionary arrangement based on the latest election returns.

However, really important decisions in Switzerland are referred to plebiscites, either on a local or national level. Voters are thus able to determine the general course for politicians to follow or to chastise their representatives if they feel that their parliamentary decisions are out of kilter. Could Helvetia be the forerunner of a future form of direct, Internet-based democracy?

Any move towards an established form of direct democracy would add a fourth power to the three traditional ones proposed by enlightenment thinkers such as John Locke and the Baron du Montesquieu. Besides the executive, legislative and judiciary branches of government, the *vox populi* would rank as equal and take part in the check&balance system that lies at the heart of modern democratic government. And this is also nothing new, harking back as it does to the Roman republic and ancient Athens, where democracy was originally invented – democratia" (δημοκρατία) literally means "rule by the people".

But what role will digitalization and networking play in this process, and what can they contribute to the principle of direct or "pure" democracy as it is called by political scientists? Well, there are after all some very practical advantages to being able to ask your constituents whenever a big decision comes up. Besides, the cost of calling an election would sink dramatically.

In addition, democracy by Internet would go a long way towards establishing a true, functioning Fourth Estate which could be independent from factional partisanship. In the past, party affiliation was the only way a citizen could make himself heard, if at all. Political influence and decision-making was channeled through the conduits of the party hierarchy, through local, regional, state and national committees and caucuses.

In an online democracy, blogs and Facebook "likes" would take the place of town hall meetings and primary elections. Online petitions pass upwards much faster than paper ones, and it doesn't take a clumsy administrative apparatus to handle them.

In such a system, parties would either cease to exist or, more likely, change their natures. By forcing the party bigwigs to actually listen to the voice of their voters and to engage in meaningful discussions about ways and means, they would not only be taken down a peg, but also empowered to actually take responsibility and act in the best interests of the majority. If not, they would be brought down at Internet speed, just like the short-lived Pirate parties in Europe.

Digital particularism

Openness and transparency seem to be the natural enemies of politicians around the world. In Germany, when Edward Snowdon revealed that the NSA had been tapping chancellor Angela Merkel's smartphone, her interior minister Hans-Peter Friedrichs told a reporter for the state-owned television network ARD that "such things should be discussed behind closed doors"; where else?

Dictators and autocrats never tire in their efforts to stem the tide of openness in the countries under their thumbs. The best example is, of course, Communist China, ably assisted by their capitalist cronies at companies like Cisco, who supplied the technology behind the "Great Chinese Firewall". This is a gigantic system of online surveillance and censorship technologies that make it exceedingly difficult (but not impossible) for Chinese citizens to surf the Web and exchange thoughts and ideas freely through Social Media. If any of these thoughts and ideas sound treasonous to the invisible censors, the thinker will probably land in jail.

But you don't have to travel afar in order to find examples of discombobulated politicians fumbling to come to grips with digitalization and networking. The European Data Protection Directive is a perfect example, stating as it does in article 25 that physically moving the personal data of European citizens outside the borders of the European Union is a criminal offense, especially if the "third country" the data is being transferred to does not abide by the same strict data protection standards practice in Europe. The most prominent such third country is, of course, the United States, where concepts of privacy and informational self-determination are either entirely different, unknown or winked at officially.

As a result, anyone who sends the names and addresses of customers, clients or employees overseas for processing, especially to the USA, could go to jail. If the world were fair, the same should apply to the heads of European spy agencies like the British GCHQ or the German MAD, who regularly swap data with their American counterparts at the NSA, as Snowdon proved. So far, no one seems to care.

That may well change quite soon. The reason lies in the rapid spread of "Cloud Computing", a principle of decentralized computing which allows corporations and government organizations to store all their information on servers located in data centers run by private companies , most of them based in the U.S. or increasingly in low-cost countries like India. Since bandwidth is no longer a problem, data can be moved around the globe with lighting speed, so that it virtually impossible for the service provider who handles them to even tell where they are at any given moment.

Besides, the nature of packet switching, on which the Internet is built, means that individual bits and pieces of data are able to find their own route to their destinations, avoiding areas of congestion or system malfunction by simply slipping around the data jam. If necessary, an e-mail from London to Paris will pass through servers in New York or Bangalore without producing any ascertainable slowdown in delivery time. And since packet switching is essentially automatic, it is extremely hard to ensure that data will not take a quick detour to another continent altogether – thus putting managers at risk of breaking national law.

Thankfully, law enforcement agencies and regulatory bodies in Europe seem to be equally clueless, so nobody gets prosecuted, at least up to now.

Large providers of cloud computing have started taking protective measures anyway. After all, who knows when the authorities will wake up? By employing technological wizardry such as "regional IP addresses" they are apparently able to guarantee their clients conformity to EU law by stopping their data from leaving the confines of the EU. This may seem to contradict the principle of packet switching and autonomous transfer protocols, but who cares? As long as my provider is willing to give it to me in writing that my data are kept safe, I as a company manager can at least rest certain that it won't be me who the police take away if something goes wrong.

In fact, the recent revelations about secret deals between the NSA and various large U.S. technology companies has proven a boon for their smaller European competitors who were slower to get off the mark in the cloud computing market. European provider such as Deutsche Telekom have started to actively market their products as clouds "made in Europe", suggesting to potential clients that they, unlike the American providers, conform to strict European data protection policies. In fact, though European spy agencies have proven just as adept as their U.S. counterparts in hacking into data networks in Europe and sharing the information gleaned that way with the authorities in the United States.

In any case, the efforts by the European Union to shut their networks off from the rest of the world bear a big potential risk creating a kind of westernized version of China's Great Firewall, not to mention repressive and intrusive systems operated by autocratic regimes in places like Russia or Saudi Arabia. After all, even so-called liberal democracies like the United States have for years been doing the same thing to root out things they disapprove of online such as child pornography or gambling, albeit with less effect than the Chinese. U.S. Internet providers are required by the authorities to block access to online casinos operating in the Caribbean, forcing these operators to move their servers around from island to island in an elaborate online shell game.

If we fail to stop governments for interfering arbitrarily with the inner workings in the Net, the result could very well be a return to something akin to the particularism that once plagued countries in central Europe like Germany and Italy which used to look like patchwork quilts made up of dozens or hundreds of independent principalities and "free cities", each one with its own laws and customs borders which stifled growth and allowed more free and unified countries like Britain or France to rise to predominance during the late Middle Ages and on into the Industrial Revolution.

We, ourselves, would hate to be forced to live in such as world where every country runs its own separate Internet, cutting off access to other cultures and shutting out people who don't follow their laws and believe in the same values they do. The term "World Wide Web" implies a network that brings all the peoples of the world together, regardless of color or creed, political or sexual orientation, taste or tradition. If that is destroyed, we would rather shut the whole show down.

Doctrines are for dummies

As we enter the Age of Digital Enlightenment, an important question we need to ask ourselves is whether a networked society still needs monolithic, ideological constructs such as the notion of the nation state in order to function.

The same goes for many other worldviews and doctrines of salvation such as those which sprang up in the 19th and 20th centuries. Communism and capitalism, fascism and anarchism, *Kodoha* in Japan, *Rexism* in Belgium, the *Acción Revolucionaria* or "Gold Shirts in Mexico, as well as a host of other universal dogmas and weltanschauung came and went, each claiming to light the only true way to true human happiness and welfare for all. Anyone who disagreed, of course, had to be expelled or purged, either by sentence of exile or death.

Now, the era of ideology is finally coming to an end, and good riddance, too! Digital societies will be based by necessity on common-sense and consensus, not because we have all suddenly and collectively become smarter, but because digital enlightenment will make it increasingly difficult for us to believe in pat solutions to improving the human condition. Instead we will need to come up with convincing arguments and systems that actually work, at least for us and those around us.

In this future world of digitally enlightened individuals the ability to think for ourselves and in real time will be crucial. Gone are the days when a Lenin or an Henry Ford could call the pace by which society marched ahead, everyone gaily swinging their arms and singing in time.

In his book, *The Wealth of Networks*[15], Yochai Benkler , the Berkman Professor of Entrepreneurial

---

[15] Yochai Benkler: The Wealth of Networks (Yale University Press) 2006

Legal Studies at Harvard Law School, predicts that networks will be the structure upon which economic progress and political organizations will be based. These structures will necessarily be in a constant state of flow, thus differing fundamentally from rigid, doctrinaire societies based in the past on some political or religious mythology or another.

Benkler sees communication and the exchange of important cultural and economic assets in the networked economy as the drivers of future growth and fortune. Classical economic theory, of course, reduces growth to the two factors originally identified by Adam Smith in the 18th century, namely capital and labor. These "hard" factors alone are considered to govern changes to national output in developed countries. Non-profit labor is not considered by classic economy to be valuable at all, since it is not seen as productive or as creating value.

Benkler points to such new developments as the Open Source movement or the online encyclopedia Wikipedia as examples of what he calls *Commons-based Peer Production* which, for him, represent a counter-concept to the currently accepted rationality and selfish behavior of *homo oeconomicus*.

The media are a good example of what he means. It used to be that printing a newspaper or running a broadcasting studio were much too expensive for most of us, so it was delegated to certain state agencies or private entrepreneurs who exercised control over their content. Today, in the age of the ubiquitous smartphone, tablet and laptop PC, the physical and economic barriers to creating and publishing content have virtually disappeared, leaving what Benkler calls *social production* as a viable means of distributing information. In this scenario, everyone is simultaneously producer and consumer of information, making us all shareholders in one of the power centers of modern society.

Those who engage in social production, Benkler thinks, see money as only one of many forms of enumeration. Others include the pride in belonging to a group of likeminded people engaged in a successful and rewarding project (such as Wikipedia) or joining in creating new digital tools for the digital age (as do the programmers who contribute code free of charge to Open Source software).

Traditional, "rational" economic models fail to take these forms of "payment" into account – and therefore miss the important point that people in the digital economy increasingly see these rewards as more desirable and motivating than a monthly paycheck.

Unlike industrial production, Peer Production leads to greater autonomy for the individual who is no longer dependent on centrally organized communication or on technology provided by employers, and who is therefore no longer dominated by them. This explains why television viewing by many people, especially among the younger set, is in decline, as well as other regulated forms of entertainment controlled by business interests.

Access to a wider range of information in turn contributes to a growing critical awareness on the part of users who will learn to discriminate between the significant and the trivial, thus increasing their own ability to think for themselves. Social and political conformity previously nurtured by press and broadcast media will play a lesser role in a networked society shaped by the yellow press and trash TV.

Similar processes are already underway in the world of business, as well. Under the headline "Where Coase went wrong", one of the authors[16] of this book once described what happens when small companies get access to unlimited digital information. One of the central tenets of Ronald Coase, the British economist and 1991 Nobel laureate, was that large corporation enjoy an intrinsic advantage due to their ability to internalize so called *transaction costs* which include the gathering and processing of market information. Production costs, Coase postulated, remain virtually identical for companies large and small, but corporate giants have a head start when it comes to finding the right product or supplier as well as other pieces of information necessary for them to conduct their business. Large companies able to internalize all their transaction costs will therefore continue to grow, while small companies will remain stunted and eventually fall by the wayside.

That size matters has long been a mantra among economists and managers, sand megalomania is part of the ´system. As a result, banks, pharmaceuticals, steel barons and car bigger and bigger manufacturers seek their salvation through bigger and bigger "mega-mergers" until they can become "global players", the

---

[16] Tim Cole: Enterprise 2020 (Hanser) 2010

ultimate goal of every industry titan, at least until the anti-trust watchdogs wake up. And even then, there are always hired political handy to provide the necessary exceptions and special permits that enable them to grow unchecked. After all, huge multinational conglomerates are a matter of national pride as well as affluence, and their state of health is often a matter of national concern.

In the stormy months of the Great Recession of 2007, the first cracks seemed to appear in this culture of gigantomania that focuses on company size as the sole determinant of success. In order to save the car manufacturer Opel, Germany was forced to mortgage its economic future to the tune of some 6.5 billion (about $8.3 billion) because, like the giant insurer AEG in the U.S. which initially cost American taxpayers at least $85 billion (most of which has since been recovered), it was deemed "too big to fail".

Not only critics of capitalism increasingly ask themselves whether big multinational corporations are really necessary, or if they couldn't be replaced by smaller, more agile companies. How many Lehmann Brothers can we afford before the whole system comes crashing down? And are conglomerates flexible enough to survive in an economic climate that is apparently marked by rapid ups and downs, by shortening cycles and the rise of superfast trading?

In other words: Do we need a "New New Economy"?

The term first surfaced in an article by Chris Anderson, the former editor of *Wired* magazine and author of *The Long Tail*[17], where he describes an economic future in which startups and small companies will replace corporate giants as the driving force for growth and progress.

The future, this implies, belongs to small companies. Thanks to electronic procurement, trade portals, intelligent software agents for e-commerce and e-business, and software-as-a-service (a kind of rental system that allows companies to turn capital costs into operating expenditures), small companies are catching up with the big boys in terms of transaction costs. The result is a level playing field, where small, nimble proponents can easily outwit and defat the cumbersome corporate dinosaurs in a classic game of survival of the fastest and fittest.

Results are already to be seen. The "Big Three" car manufacturers in America, for instance, have begun to outsource entire components such as power trains or suspension systems to small (or at least: smaller) contractors and in the process reinventing themselves as "networked corporations".

And another factor is helping small companies succeed: cloud computing, which essentially means outsourcing your IT infrastructure, leaving small companies to concentrate on what they do best, namely their business. Instead of keeping lots of capital and manpower wound up in maintaining computer systems, networks and data storage, they can simply rent as they go, adding capacity when necessary and

---

[17] Chris Anderson: The Long Tail: Why the Future of Business Is Selling Less of More (Hachette Books) 2008

booking everything under operating costs. This in turn makes them even more flexible and ready to expand into new areas or grow their existing capabilities.

The future off intelligence

Vinton Cerf, one of the inventors of TCP/IP, is fond of talking about what he calls "network intelligence". By that he means the makeover of IT landscapes triggered by digital networks and leading to rapid and fundamental changes in technology.

Computer experts often refer to this process as "Digital Transformation", and there is no reason to believe that it will only have an effect in technology. In fact, digital transformation is going on as we speak in humans who are the users and beneficiaries of this digital technology. We ourselves are changing the way we communicate, enjoy ourselves, do business and organize society. How's that for "digital transformation"?

Digitalization and networking may actually spawn the development of a new, superior form of human intelligence, which we will henceforth call *communicative intelligence*. The collective IQ of a network community, we believe, is greater than the total sum of the IQs of each of its members. In fact it is entirely possible that this collective IQ will behave according to the *network effect* first postulated by John Metcalf n 1993 which states that the benefit of networks will grow exponentially to the number of users.

Communicative intelligence will shape the way markets behave, according to the *Cluetrain Manifesto*, the unofficial bible of online marketers published in 1991 by Rick Levine and his colleagues Doc Searls and David Weinberger. The book consists of a collection of 95 theses that describe how the relationship between vendors and customers will evolve as the Internet and the "New Economy" cause tectonic shifts in the balance between the two.

Markets today are conversations, the Cluetrain Manifesto states. This is easily apparent when we think about how the act of purchasing things has changed over the past few years from a linear process (you saw something, you went to the store and bought it; vendors could focus their marketing efforts almost exclusively on the point of sale) to a networked one. Today, we see or hear about things on Facebook or Twitter; we listen around to find out what other customers have to say; we look up prices on comparison portals and buy from however offers us the best deal. And the conversation continues even after we have bought the product, because we immediately start sharing our experiences and impressions with others on the Social Web.

In marketing-speak, this is now known as the "customer journey", the process itself is often referred to as "social shopping". The vendor, it turns out, is largely reduced to the role of spectator in this game, at least until the customer actually clicks on the "buy" button. All the vendor can hope to achieve is to be allowed to take part in the conversation, but even that is not at all certain.

So what will the future bring? What will the results of communicative intelligence and digital networks on individuals and on society as a whole be? Is homo sapiens really evolution's last word, or are we on the brink of the next step in the development of human intelligence, this time triggered by technology and our use of more and more intelligent tools? And more to the point: How will we know if we have moved to a new, a higher form of human intelligence?

Maybe what we need is a kind of "social Turing test". Alan Turing, many will remember, was a British computer scientist who suggested a way of testing machines for intelligence long before there were computers in the modern sense of the word. Turning spent most of World War II unraveling the secrets of the German army's "Enigma" coding devices and later committed suicide after his homosexuality came out which carried both a social stigma and the threat of legal persecution.

Turing envisioned a test setup consisting of a human interrogator connected somehow (since it was a though experiment, details didn't matter) to another human being and the mechanical device being tested, which were both sitting behind a screen. The tester would ask both of them as many questions as necessary in order to determine which the machine was. If the tester failed to find any difference, it would be assumed that the machine was demonstrating a degree of intelligence equal to that of a human being.

Incidentally, no computer program has ever been able to pass the turning Test, and not for lack of trying, either. Every year, a competition is held for a prize established in 1990 by the American inventor Hugh Loebner. The $100,000 for the first algorithm to convincingly exhibit human-like intelligence has yet to be claimed.

The collective intelligence of a society will be measured in future by the amount of information it can amass and make available. The society with the greatest store of accessible information will be superior to other, less informed societies and political systems. The information needs to be "autonomous", for instance because it is stored in the Cloud and is therefore available to anyone who wants it. If we achieve that, then we will truly be able to say of ourselves as a society that we are equipped to deal with the challenges digitalization and networking present us with by using information in a productive and creative way.

Chapter 11

Think For Yourself!

Think For Yourself!

Fundamental change doesn't just „happen", even in times of digitalization and networking. Masses of people will need to pitch in first in order to stir up social change. In the process, they will need to create a whole new set of categories to describe what is happening; only then will they be able to start imagining the new. This means redefining familiar terms and concepts and interpreting them in new political and philosophical ways. The "unthinkable" is only so because we lack the words to think it.

But writing a new dictionary isn't enough; we need to fill these terms with new meaning denoting new values that are so compelling that people will invest their time, their treasure and possibly much more to achieve them.

Following the brutal murder of French journalists working for the satirical magazine "Charlie Hebdo", people all over the world united in condemning the attack on the right of free expression. Some reminded us that this right was fought for by people with the courage to risk their very lives for an idea born from Rational Enlightenment. We will need to harness similar powers of convictions if we can ever hope to reach Digital Enlightenment.

Whenever people discuss political and philosophical ethics, they first have to agree on a common terminology. Only then can they start describing a new moral standpoint in order to elevate it to the level of a universal ethical principle. Immanuel Kant had a similar problem in his day, and he went so far as to demand that the terminology used in a discussion of ethics must actually contain the elements of that ethic in order to lead to relevant results.

Ethics never stands alone; it must emerge from a discourse between conflicting systems if it is to stand the test of time. The history of philosophy is full of examples, from Aristotle's *Nicomachean Ethics* to Kant's *Categorical Imperative* which contains his "Groundwork of the Metaphysics of Morals", as well as the ancient Indian *Bhagavad Gita* or, in modern times, Jürgen Habermas and his *Ethic of Discussion*. Without dialogue, ethics are nothing.

By the same measure, any future ethical system, in order to be acceptable to society in general, will require wide consensus about the values it is to represent. A proper terminology is crucial here if we all are to clearly understood what we are talking about. Any yes, a certain amount of controversy is to be expected, since different groups represent different interests. That is why Kant took such care in defining the terms he used to describe rational enlightenment.

New words for a new system of ethics

An ethical system of philosophy that is not based on a common terminology and full understanding of first principles can only be pontifical and pastoral. After all, such a system cannot call upon a deity for its own justification.

Kant is quite explicit when he describes this connection between ethics and its own terminology, between ideas and words. These, he said, must already include the ethical concepts they describe. And he insisted that we must be able to think for ourselves, summoning up the courage to use our own powers of reasoning, and this not only in the comfort of our homes but openly and publically.

In attempting to establish ground rules for a new Digital Enlightenment, must aspire to these same lofty ideals. Before setting out in search of new values to guide us, we first need to agree on the terminology we intend to use. For Kant, terms like "reason", "courage" and "common sense" were his word-tools of choice. Today, we need new ones (although the some of those used by Kant are still valid and important).

We do not deceive ourselves that this book can provide a comprehensive dictionary of the new language of Digital Enlightenment. That will require long debates. If anything, this book can lead to a more focused discussion on the part of those concerned --- which means everybody, regardless of sex, creed or color and especially political leanings and cultural tastes. The sheer diversity of opinions and beliefs within the digital community will necessarily lead to the creation of a rich new vocabulary for the Digital Age.

All will be made clear

What we can do, however, is to demonstrate how certain terms can change their meaning to fit the times, so to speak. We will try to show how political and social values can take on new technical, mathematical or scientific meanings under the influence of digitalization and networking, depending on context.

A good example is the word "transparent", an ambiguous term if there ever was one. Taken in the context of IT and software development, transparency describes a process that is either intuitive to use of even unnoticed, running comfortably in the background. On the other hand, when the term "transparent" is applied to a political, social or philosophical system, it implies principles that are clearly understood and obvious to everybody.

Users of a "transparent" social network, for instance, don't need to know how it works to use or be a part of it, and it will continue to function without supervision. The exact opposite applies when you use "transparent" in a political, social or philosophical context, where it is a desirable property because it allows everyone to understand what is going on and where we're heading.

In any case, in a culture marked by sharing and communicating, transparency is a value in itself because without it no serious and meaningful discussion is possible.

A simple word as "transparency", used in this way, is of course in danger of suffering semantic overload. However, it is debatable that the reverse is happening. In fact we feel that the term describes quite pointedly what is going on in various social contexts.

In any case, the term clearly refers to itself and its meaning as well as to a value it is closely related to, so it complies with Kant's definition. A functioning network that is simple to navigate and practical to use since it allows us to communicate without us concerning ourselves with its upkeep and maintenance is for all intents and purposes invisible. And we can expect such a network to become popular in a very short time as more and more people discover how easy it is to use, thus leading to the "network effect" described earlier. It therefore will become increasingly valuable, both socially and economically. As a result, the original technical meaning of the word "transparent" takes on a new social dimension.

But even this new sociopolitical dimension, it turns out, can be ambiguous, too. For instance, when we talk about "transparent markets", we imply these are "free" and function according to clearly defined rules. The opposite would be a "regulated" market where the rules are set up by someone else. The strictness or arbitrariness of these rules will depend on the political instincts of the regulators instead of on how market usually work; no wonder they are so beloved by conservatives and others who pay only lip service to market liberalism.

In the context of political decision making, transparency takes on yet another meaning. Times were when liberals were fond of it, but it has been largely abandoned by them in the attempts to appeal to conservative voters, so it was taken up by the Pirate parties instead. "Transparency" in this context describes a political process which is open and aboveboard, and where decisions are no longer reached in smoke-filled rooms but are the result of informed debate between citizens of all stripes.

A transparent political system, therefore, is one which functions in full view of the voter according to rules agreed upon by all and bent on achieving goals everyone supports. Transparency therefore can ennoble a political system and raise it above other, less transparent systems.

The future is open

A second key term in our new vocabulary of Digital Enlightenment is openness, which is closely related to but not identical with transparency, and admittedly there is a risk that less enlightened minds will confuse the two. We will have to live with that.

Systems like the Open Source movement, which mandates that programmers publish their code for free, or the Open Data initiative which is based on the principle of free access to all data, have been part of online culture and economy from the very beginning.

The antonyms of openness are "closed" or "proprietary" solutions and services which are typical for corporate cultures and which are based on such traditional concepts as copyright and intellectual property. In a capitalist system, protection of property is a top economic priority, and it has therefore been transformed into a moral imperative. Immaterial property, the reasoning goes, also leads to material gain and so should enjoy similar protections. In fact, though, in the digital sphere open systems have long proving themselves both more effective and more profitable, at least in the long run.

Just a few years ago, "open" systems were considered anathema by most computer scientists and software developers. This started to change only once the Open Source movement began to demonstrate its technical and economic superiority. The reason lies in the very nature of openness: Once programmers are allowed to reuse existing code created for some other project instead of replicating the efforts of their peers, they can use their talents to create something even better. The only requirement is that they agree to hand what they have created back to the community by agreeing to its use under the terms of "Creative Commons" agreement. That way, software can be developed much faster. In addition, the quality of the software will usually be higher because of the process of "peer review", which is also part of the Open Source concept, performed for free by the online community as a form of self-regulation.

The business benefits of such a system are obvious. Programmers and entrepreneurs get to use the code for free, speeding up the development process, and leaving them free to concentrate their efforts (and their capital) on providing service and support for the products they have created.

Richard Stallman, a programmer and activist, was one of the first to push the open source approach as opposed to proprietary software development. Stallman became one of the founders of the Free Software Foundation, an association devoted to the concept of "freeware": programs that are often, but not always, simply given away (donations are welcome!).

One of Stallman's pet propositions was the "GNU Project", which stand for „GNU is Not Unix". Stallman and his friends wanted to demonstrate that you can develop an entire computer operation system from scratch using freely available digital "building blocks".

GNU was an early forerunner of Linux, the Open Source operating system for personal computers that has since been installed on at least 67 million PS around the world, according to *Linuxcounter*, itself another Open Source project.

Stallman was quite a radical, and his aggressive attacks on the "commercialism" practiced by his fellow software developers eventually led to him being ostracized by the community; one of the most poignant dramas in the history of computing. His place was taken by Linus Thorvald, the author of Linux, who proved to be more politic in his dealings with other software nerds, so he soon became the poster child of the Open Source movement. It is largely due to the personal charm and open-mindedness (sic!) of this Finnish-American software engineer that Open Source is not only accepted by most computer professionals today, but also has become integrally linked to the globally networked digital community.

Openness is undoubtedly one of the keys to accelerated innovation, not only in the technical sense, socially, as well. Openness not only helps us to assimilate latest technologies, but it exposes us to foreign cultures, too, turning it into an important value, and one without which Digital Enlightenment would be impossible to achieve.

Autonomy as a system

The third term we would like to add to the growing dictionary of Digital Enlightenment is "self-regulation". In cybernetics, systems that regulate themselves are widespread; in fact the Internet itself is probably the best example of an "autonomous", self-regulating system since it was designed to let individual "packets" of information seek their own way to their destination; they then reassemble themselves and can be displayed in their original context. Any attempt at interfering with the flow of information in such an automated system, for instance censorship, is routinely treated as just another obstacle – something the system is expressly designed to work around.

The principle applies as well to "non-technical" networks. "Hippie Havens – land communes fostered by members of the counter culture in California and elsewhere in the 70ies and 80ies – are a good example. "The Farm", a community founded in Tennessee by hippies from San Francisco, is the oldest existing of these, having been established in 1971. At its height it boasted 1,500 members and attracted such celebrities as Walter Cronkite and Phil Donahue. Today, this oldest hippie commune is still going strong, as a recent documentary by *abc News* proved.

Social Networks and community projects like Wikipedia are other well-known instances of self-regulating (and often self-policing) systems. Taking this idea a step further, traditional non-technical communities such as hunters and gatherers, agrarian village communities and many spiritual or religious groups can be seen as autonomous social networks since they aim less at improving material living conditions but instead maintain their cohesiveness through the sharing of resources and common values based on shared experiences. Thanks to digitalization and networking, it is easily possible to set up similar communities regardless of physical distance or political borders and even without users being required to speak a common language. Such networks do much more than just allow members to communicate; they can exchange opinions and feelings, set common goals and define common values.

Terms like self-regulation, openness and transparency are not neutral by any means; quite the contrary. The more widespread they become, the more their intrinsic value increases and the more they give us room to think for ourselves, which is a necessary condition for Digital Enlightenment. Their effects are huge since they provide new channels of communication and collaboration as well as the ability to share and to innovate. All this makes them true agents of enlightenment.

Digitalization and networking, we believe, are currently reshaping our lives, as well as our understanding of ourselves and others. They are even starting to change how we think and feel, thus making equipping us to face the future. Digital Enlightenment turns us from helpless and hapless victims of change into active participants, from spectators to players.

There is a still lot of work to be done. For instance, we need to define the new terms more precisely so that we can better understand them. Eventually, we believe, their inherent values will become obvious to all or most of us. But this doesn't mean we have to create a fixed and immutable "canon of values", like church dogmas. The authors are not seeking to establish a new religion. We simply want to echo Kant's dictum, one which he himself borrowed from the Roman philosopher Horace a couple of millennia earlier, namely: "sapere aude!" --- "dare to know!"

We need you, gentle reader, is we are to achieve that goal. Not because we are afraid or incapable of doing it ourselves. Only by pooling our efforts and working together in new ways using new technologies can we hope to contribute collectively to the establishment of Digital Enlightenment. In fact, this book is not and was never intended to be end in itself. Hopefully, however, it will contribute to an ongoing, informed and fruitful dialog about words and values that can lead us forward. We'll keep you posted!

Afterword:

How we wrote this book

This book took more than 20 years for us to write. Ossi Urchs and I go back even further, all the way to the early 80ies when we both wrote for the German edition of *Playboy* which then was one of the most successful magazines in Germany (as indeed it was in the U.S. as well). We both belonged to the small, select group of "Edelfedern"; eloquent essayists that Fred Baumgaertl, the legendary editor-in-chief, surrounded himself with and whom he allowed to write about almost anything they liked.

Ossi was seriously into technology in general and virtual reality in particular, which was how he got to know Jaron Lanier so well. I had done a stint in hifi audio journalism and was getting involved in video gaming. Later, in the early 90ies, I became head of the "multimedia editorial group" at a large German publisher called *Motor Presse*, and I asked Ossi to write me an article about the mysterious new phenomenon people were staring to talk about called the "Internet".

Ossi wrote a wonderful piece that focused mainly on the rock band "Grateful Dead" and their fans, most of whom were getting on in years by then, and who were wont to communicate with each other via computer through so-called "bulletin boards" that were cropping up all over cyberspace. In them, the aging hippies would exchange messages about their kids and grandkids, about concerts they had been to and where will we meet next time? And they also shared audio recordings of concerts by the band (with their express permission) through something called "FTP".

I immediately got my own Internet connection and started using "FTP", writing e-mails and doing something called "Gopher" which allowed you to see the folders on someone else's computer. That was all there was backing then. And then the World Wide Web came along, and neither Ossi nor I ever looked back.

In 1995, I began to publish what I then called my "Online Diary" (the term "blog" hadn't been invented yet), and Ossi founded a company that built Websites for companies that wanted to get a foot in the door of this exciting new medium.

We both attended CeBIT every year, the giant German computer fair in Hanover, and we started to form the habit of sitting down at least once to a fine dinner at our favorite Italian restaurant, "Roma", where Lino, the owner, would serve us "a few noodles", and "a little wine" which made us feel like we were really sitting at Piazza Navona or looking out on Mount Vesuvius.

At the time, Ossi was living in the future; quite literally: He had been hired by Philip Morris to act as their "Minister for Tomorrow", displaying his long red dreadlocks on countless billboards and in TV ads and challenging the rest of us to follow him to such thrilling places as Silicon Valley or Manhattan's "Silicon Alley" to find out what the future held in store.

Every year at CeBIT, while stuffing ourselves with pasta and scampi, we would talk about all the exciting things that we had learned in the course of the year. These conversations usually went on and on until Lino told us he needed to close up, and in the course of our talks we solved most of the most pressing problems of technology and of humanity. Unfortunately, given the quality and quantity of the wine Lino served us, we were never able to remember the solutions we had found with any degree of clarity. We only knew that they were great, and that neither of us would have been able to dream them up on his own; the whole, it seemed, was always greater them the sum of its parts, as Aristotle was fond of saying.

The idea of turning these ideas into a book came to us early on, but we simply didn't have the time to sit down and start writing. Ossi was busy with his Internet company, with TV work and giving speeches all over Europe. I had recently written my first book, "Success Factor Internet", which brought me a great number of speaking engagements, too. Our paths sometimes crossed on some stage or the other, and of course we still had our annual evening at CeBIT to look forward to, so the dialogue went on.

My editor at the company that published most of my books, Martin Janik, was the one who finally brought us together. He was very keen on the idea of publishing a book by both of us. Ossi held off initially because he was too busy, but I finally managed to convince him that if two people write a book that means each only does half the work.

How wrong I was! It turned out that creating a book that has two authors is a tricky and time-consuming proposition. There are really only two ways you can go: Either each of you writes half of the book alone, and later someone tries to figure out how to combine two very different styles of thinking and writing into a single, humongous volume; or you hark back, as we did, to our tradition of Platonic dialogue and try to create a book by discussing certain subjects, recording the conversation, creating a transcript and sending the results back and forth a few times by e-mail for editing until you finally agree on a version you both can live with.

That's what we did, and the result is a curious and, at least for me, haunting blend of two very different backgrounds. Ossi was a student of the "Frankfurt School" of modern philosophy and media theory, as well as an avid student of ancient Indian philosophy which he studied under a string of gurus (a word that means 'teacher' in Sanskrit). I am an American expatriate who, despite having lived in Europe for almost 50 years, has always retained his quintessential playful American fascination for technology in general, as well as a bright and cheerful expectation of what the future has in store for us.

We solved the problem by arranging to meet for days on end, first in Ossi's cottage in the Spessart mountains north of Frankfurt, and later at my house in the Lungau, a remote valley in the Austrian Alps south of Salzburg, to spend days sitting around and talking, with the recording device running all the time.

Our "Spessart Talks" were full of vim and high spirits, and we both looked forward to completing our task in record time. But then one day Ossi called me up and told me that the doctors had found something growing n his brain that was looking distinctly worrying. After a series of examinations it turned out to be much worse: He was also suffering from lung cancer, and he would require extensive chemotherapy. There was no telling when – or if – he would be able to get back to work.

This was bad news in all sorts of ways. In terms of the book, it meant that some of the most important chapters were yet unwritten, for instance chapters 3 ("Thinking in real time") and 11 ("Think for yourself!") with its call for an informed social discussion about the terminology and values of Digital Enlightenment.

I called Martin Janik and he reacted wonderfully: Ossi should please concentrate all his energies on getting well again, and of course the contract we both had signed would remain in effect, regardless. As for our deadline, hey, just see how it goes!

It took almost half a year, but Ossi finally emerged from hospital apparently as healthy and eager to go on with the writing as ever. Okay, his red dreadlocks were gone, tucked away in a shoebox awaiting the day they would submerged in the waters of Mother Ganga or some other river. But his head was as full as ever with ideas waiting to be discussed, recorded and eventually set down on paper.

Our conversations thereafter produced texts that for me belong among the most clearly and closely argued and logically derived I have ever read. Neither Ossi nor I could have ever achieved such results on our own, and reading over them again today I sometimes feel my skin crawl. Yes, Ossi, this was a true example of thinking going on in real time, as you insisted again and again we all should learn to do.

In a certain sense, Ossi and I wrote this book for us, and I find that is usually the best kind of book. It may sound self-centered, but in fact writing it was not only fun, but also extremely stimulating, like a ride on some kind of intellectual roller coaster; one you never knew where it would finally take you. We simply had to get all this out of our systems, and if somebody else enjoys it too, then that's fine, but if anyone doesn't like it then tough luck. This is, after all, a book about the dangers of cultural pessimism and the deep-seated technophobia we find in certain parts of society, so it almost goes without saying that we stepped on a few toes.

*

When I originally wrote this afterword, Ossi seemed to be making a full recovery. Alas, that proved a short-lived hope. Less than half a year after the German version was published, he passed away and was buried near his home in Offenbach, a place he always jokingly referred to as the "South Bronx of Frankfurt".

It took us 20 years, but I personally wouldn't miss them for anything. But I sure will miss those conversations. So long, Ossi! See you in in the future, maybe.

St. Michael im Lungau/Salzburg, February 2015

Books that inspired us

Anderson, Chris The Long Tail – der lange Schwanz. Nischenprodukte statt Massenmarkt – Das Geschäft der Zukunft, München (Hanser) 2007

Aristoteles: Nikomachische Ethik, Hamburg (Felix Meiner Verlag) 1972

Bardeau, Frederic und Nicolas Danet: Anonymous – Von der Spaßbewegung zur Medienguerilla, Münster (Unrast Verlag) 2012

Benkler, Yochai: The Wealth of Networks, New Haven (YaleUniversity Press) 2006

Berners-Lee, Sir Tim: Der Web Report, Berlin (Econ) 1999

Berners-Lee, Tim: Weaving the Web, New York (HarperCollins) 1999

Bomhard, Sebastian von: World Wide Was?, München (SpaceNet) 2009

Braun, Gabriele (Hrsg.): Digitaler Dialog, Waghäusel (marketingBÖRSE)2012

Bunz, Mercedes: Die stille Revolution, Frankfurt (edition unseld) 2012

Burton, Richard: Anatomie der Melancholie, Dieterlch'sche Verlagsbuchhandlung; Auflage: 4. (2001)

Canter, Laurence und Siegel, Martha: Profit im Internet, Düsseldorf (Metropolitan)1995

Carrol, Lewis: Through the Looking Glass, London (Macmillan) 1896

Cole, Tim: Unternehmen 2020 – das Internet war erst der Anfang, München (Hanser) 2010

Dobelli, Rolf: Die Kunst des klaren Denkens, München (Hanser) 2011

Dueck, GunterAbschied vom Homo oeconomicus, Frankfurt (Eichborn) 2008

Friedman, Thomas L.: The World is Flat, New York (Farrar, Strauss and Giroux)2005

Gleick, James: The Information, Toronto (Random House Canada) 2011

Habermas, Jürgen: Erläuterungen zur Diskursethik. Frankfurt am Main (Suhrkamp) 1991

Kant, Immanuel: Kritik der praktischen Vernunft, Leipzig (Felix Meiner Verlag) 1929

Kant, Immanuel: Was ist Aufklärung?, in: Berlinische Monatsschrift, Dezember-Heft 1784, p. 481 bis 494

Kelly, Kevin: What Technology Wants, New York (Penguin) 2010

Kurzweil, Ray: The Age of Spiritual Machines - When Computers Exceed Human Intelligence, New York (Viking) 1999

Kurzweil, Ray: Homo S@piens, Leben im 21. Jahrhundert – was bleibt vom Menschen? Kiepenheuer & Witsch (1999)

Lanier, Jaron: You Are Not a Gadget, New York (Peguin) 2011

Lanier, Jaron: Gadget: Warum die Zukunft uns noch braucht, Frankfurt/Main (Suhrkamp) 2010

Lessig, Lawrence: Code 2.0, New York (Perseus) 2006

Levine, Rick et al.: The Cluetrain Manifesto – The End of Business as Usual; Cambridge, Ms. (Perseus Books) 2000; Online unter: http://www.cluetrain.com/

Lobo, Sascha, Katrin Passig: Internet – Segen oder Fluch, Berlin, (Rowohlt) 2012

Lovink, Geert: das Halbwegs Soziale – Eine Krik der Vernetzungskultur, Bielefeld (transcript verlag) 2012

McLuhan, Marshall: The Gutenberg Galaxy, Toronto (University Press) 1962

Metzinger, Thomas: Subjekt und Selbstmodell, Paderborn (Mentis-Verlag) 1999

Metzinger, Thomas: Der Ego Tunnel, Berlin (BV Berlin Verlag), 2009

Morozov, Evgeny: To Save Everything, Click Here, New York (Public Affairs) 2013

Nefiodow, Leo A.: Der sechste Kondratieff, Sankt Augustin (Rhein-Sieg) 1996

Schirrmacher, Frank.: Payback, München (Karl Blessing) 2009

Schmidt, Eric und Jared Cohen: Die Vernetzung der Welt – Ein Blick n unsere Zukunft, Reinbek (Rowohlt) 2013

Schumpeter, Josef: Kapitalismus, Sozialismus und Demokratie, Dieterich'sche Verlagsbuchhandlung (2001)

Schwartz, Evan: Webonomics, New York (Broadway) 1997

Schwarz, Torsten: Leitfaden Online-Marketing, Waghäusel (onlineBÖRSE) 2007

Searls, Doc: The Intention Economy, Boston (Harvard Business Review Press) 2012

Seelmann-Holzmann, Hanne: Cultural Intelligence, Wiesbaden (GWV Fachverlag) 2010

Siebel, Thomas M.: Cyber Rules, New York (Doubleday) 2000

Siegele, Ludwig und Zepelin, Joachim: Matrix der Welt, Frankfurt (Campus) 2009

Sloterdijk, Peter: Du musst dein Leben ändern, Frankfurt/Main (Suhrkamp) 2009

Sloterdijk, Peter: Sphären I – III; Frankfurt/Main (Suhrkamp) 1998

Sloterdijk, Peter: Weltfremdheit, Frankfurt/Main (Suhrkamp) 1993

Stöcker, Christian: Nerd Attack!, München (DVA) 2011

Tripura Rahasya (Die geheime Botschaft der Göttin Tripura), Deutsche Ausgabe, Interlaken (Ansata) 1986

Turkle, Sherry: Verloren unter 100 Freunden, München (Riemann) 2012

v. Gehlen, Dirk: Mashup – Lob der Kopie, Frankfurt/Main (Suhrkamp) 2011

v.Samsonow, Elisabeth: Egon Schiele: Ich bin die Vielen, Wien (Passagen Verlag) 2010

Wiener, Norbert: Cybernetics or Control and Communication in the Animal and the Machine, Boston (MIT Press) 1948

Wilson, Robert A.: Die Illuminati-Papiere, Reinbek (Rowohlt) 1983

Index

"@lusches 9

3D printers 139
9/11 204
Accountemps 115
ACTA 209
Adam Smith 229
ADHD 92
Advaita Vedanta 47
AEG 232
Aeschylus 81
Afghan Citadel 130
Age of Acceleration 33
Alan Turing 235
Albrecht Duerer 180
Alzheimer 90
Amazon 101, 175
Anders Behring Breivik 83
Anders Breivick 166
Andreas Zielcke 201
Andrew H. Knoll 96
Angela Merkel 65, 223
anonymity 166
Anonymous 180
Anonymous (movement) 218
AOL 194
Apple 185
Apple Watch 27
Arab Spring 108
ARD 99
Aristotle 240
Atman 47
Atos 129
autonomous 247
avatar 168
Barbara Prammer 221
Battle for Seattle 219
Ben Kaufman 144
Bhagavad Gita 240
Big Three 233
Bitkom 120
black holes in cyberspace 177

Blackberry 33, 128
Bloggers 193
Brain Activity Map Project 90
burqa 165
Categorical Imperative 240
CeBIT 252
Centers for Disease Control 94
CERN 12
Chaos Computer Club 170
Charles Darwin 95
Charles Kettering 215
Charlie Hebdo 239
Chris Anderson 232
Christopher deCarmes 89
Christopher Lang 76
Cisco 224
Clay Shirky 117
clean desk policies 135
Cloud Computing 225
Cluetrain Manifesto 234
common sense 241
Commons-based Peer Production 229
communicative intelligence 234
CompuServe 100
Computer Right 171
Conrad Zuse 131
CopWatch 208
Copwatch Paris Nord I-D-F 208
copyright 179, 186
Courtney Love 183
CPA 91
crackberries 148
Craig Venter 90
Creative Commons 245
Crowdsourcing 178
customer journey 235
cut&paste 192
data protection 226
Dattatreya 47
David Hilbert 209
David Weinberger 234

Index

Defense Distributed 139
de-industrialization 138
delayed gratification 106
demanding systematic change. But what is Digital 217
democracy 222
Deutsche Telekom 226
Digital Acceleration 128
Digital Bedouins 147
Digital Natives 68
Digital Omerta 160
Digital Transformation 15, 234
direct democracy 221
Dirk von Gehlen 66
Doc Searls 234
Dreaming Paths 45
Edelman 194
Edward de Vere 180
Edward Snowdon 204, 209, 223
Emilio Mordini 153
Enclosure Movement 145
end of history illusion 96
Enigma 235
Eric Schmidt 14
Ernst August Wagner 155
Ernst Jünger 81
ERSP 195
Ethernet 52
Euromold 139
European Data Protection Directive 224
European Data Retention Policy 159
European Union 224
Facebook 53, 119, 161, 223, 235
Facebook Society 109
factory of tomorrow 140
Farewell Address 219
FBI 218
Foursquare 118
Fox News 191

Frank Rieger 170
Frank Schirrmacher 19, 36, 65, 76
Frankfurt School 254
Fraunhofer Institute 132
Fred Baumgaertl 251
Frédéric Bardeau 218
freeters 134
Fringe 89
FTP 252
GCHQ 224
General Electric 140
Generation Now! 108
George Dyson 210
George Holliday 207
George Washi 219
German High Court 170
Gilgamesh 192
glass customers 177
Globalization 3.0 131
GNU Project 245
Gold Shirts 228
Google 14, 176, 193
Google Glass 27
Google+ 109
Gopher 252
Gordon Dahl 84
Gordon E. Moore 25
Gottfried Wilhelm Leibnitz 202
Gowalla 159
Grandmaster Flash 188
Grateful Dead 251
Great Chinese Firewall 224
Great Earthquake of Lisbon 202
Great Recession 232
Grisham, John 176
guru 47
Gutenberg Research College 66
hacktivists 139
Hans Eichel 220
Hans-Peter Friedrichs 223

Harry Reid 177
Heinrich von Kleist 49
Herbal Group 195
Herbert Freudenberger 127
Hilmar Hoffmann 220
homo ecconomicus 229
homo mobilis 148
Horace 249
HTML 13
HTTP 13
Huffington Post 194
Hugh Loebner 236
Husni Mubarak 216
HyperCard 12
IATA 145
IBM 33
Identity Women 167
Immanuel Kant 37, 59, 240
Industrial Big Data 142
Industrial Revolution 137, 227
Industrie 4.0 137
infantilization 107
ING Dipa Ipsos 114
instant gratification 106
Intel Corporation 25
intellectual property 179, 184
Internet Speed 7
iPhone 33, 54
Isaac Newton 29
Jacques Soulillou 189
James Cameron 168
James Eagan Holmes 83
James Metcalf 52
Jaron Lanier 56, 76, 81, 105, 178, 251
Jean Michel Jarre 188
Jeff Bezos 102
Jimi Hendrix 45
Jody Markopoulos 141
Johannes Guttenberg 180
John Archibald Wheeler 75
John Gage 171
John Naughton 182

John von Neumann 85
Joichi Ito 117
Jon Stewart 30, 198
Jost Smiers 187
Judith Ramey 130
Jürgen Habermas 240
Justin Amash 209
Juvenal 203
Kaliya Hamlin 167
Karen Renaud 130
Kevin Kelley 50
Kevin Kelly 73
Kim Cameron 155
Kodoha 228
Konosuke Matsushita 20
Kraftwerk 188
Kurt Cobain 183
Lamar S. Smith 176
Larry Ellison 128
Lawrence Durell 19
Learnstuff 115
Leigh Shaw-Taylor 144
Leviathan 220
Lewis Carrol 7
libertarianism 164
Linda A. Goldstein 196
Linda Stone 77, 91
Linus Thorvald 246
Linux 246
Linuxcounter 246
Lisbon 201
Lungau 254
MAD 224
Madonna 183
Manfred Spitzer 19
manufactura 77
Marissa Mayer 127
Marshall McLuhan 51
Martin Janik 253
Maximilian I 180
Maya 48
McKinsey 142
Mercedes Bunz 71
Metcalf's Law 52

Index

Microsoft 155
MiracleBurn 195
MIT Media Lab 146, 193
MMORPG 168
Modular Transverse Matrix 143
Moksha 48
Moore's Law 87
Motor Presse 251
multitasking 65
NARC 195
National Bureau for Economic Research 84
neophiles 95
network effect 53, 243
network intelligence 233
neuphobes 95
New Economy 234
New New Economy 232
Nicholas Negroponte 146
Nick Carr 19, 81
Nicolai Kondratieff 87
Nicolas Danet 218
Nicomachean Ethics 240
Niels Bohr 75
Nokia 32
Norbert Bolz 21, 84
Nova Spivack 158
NSA 36, 64, 158, 209, 218, 224
NSAgate 203
Occupy movement 108
Offenbach 257
omerta 160
Omneuron 89
Opel 232
Open Data 244
Open Source 55, 178, 244
openness 210, 244
Otohime 158
Patrick O'Brian 127
Paul Saffo 148
peer production 230
peer-2-peer 55
Peter Principle 56

Peter Sloterdijk 35, 63
Phil Donahue 247
PIPA 176
Pirate Parties 108, 217, 223, 243
plagiarism 190
poetry slam 184
political parties 220
Popol Vuh 192
Post-Privacy 158
prana 46
privacy 156
Prostalex Plus 195
Pulchinella's Secret 154
Puluga Saga 192
PwC 195
Pythagoras 48
Quinnipiac University 205
Quirky 144
Radicali Group 82
Ralph Manning 203
Rapneck Ossi 189
Ray Kurzweil 10, 85
real name policy 166
real time 63
Recep Erdoğan 215
recontextualizing 189
Rexism 228
Richard Alan Friedman 93
Richard Allen Posner 185
Richard Stallman 245
Rick Levine 234
Rilke 63
Robert Anton Wilson 23, 95
Robert Burton 20
Robert Spitzer 92
Roby Massarotto 208
Rodney King 208
Roland Emmerich 180
Rolling Stones 183
Roma 252
Ronald Coase 231
Roya Maboob 130
Rumpelstiltskin Effect 164

same-day delivery 102
sapere aude 249
Sapere aude 59
Sascha Lobo 147
Scientology 218
Scot Joplin 45
Seat 143
Second Life 168
self-regulation 247
Shakespeare 180
Shankara Charya 47
sharing 185
Siemens 33
Silicon Valley 253
Singularity 85
Škoda 143
social engineering 120
social media 121
Social Media 224
social media guidelines 122
Social Networks 247
social production 229
Socrates 19, 81
Sony 33, 218
SOPA 176
Sound Princess 158
Spessart 254
Spider Woman 45
Stationers' Company 180
Statues of Anne 180
Stefano Della Vigna 84
Steven Jobs 128
Steven Levy 175
Steven Pinker 23
Sun Microsystems 171
supercycles 88
Supreme Court 186
swarm intelligence 56
Switzerland 219
T.S. Eliot 188
Tahir Square 216
Tantra 46
TAPs 97
TCP/IP 13, 233

technicum 73
tetractys 48
The Farm 247
The Long Tail 232
Theodore Kaczyinksi 88
Thierry Breton 129
third country 224
Thomas Hobbes 220
Thomas L. Friedmann 131
Thomas Metzinger 66
Tim Berners-Lee 12, 43
Tom Friedman 203
transaction costs 231
transparency 210, 242
Tripura Rahasya 46
TRPs 97
Twitter 8, 120, 215, 235
United Nations 170
Universal Declaration of Human Rights 170
Universal Rights of Man 208
Upanishad 48
Urban Nutrition 195
Ursula von der Leyen 164
Vedanta 46
Vinton Cerf 15, 233
virtuality 75
Vishnu 168
Volkswagen 143
vox populi 222
wa 21
Walter Cronkite 247
Walter Mitty 164
whistleblower 205
Whole Earth Catalog 73
WikiLeaks 36, 153
Wikipedia 176, 229, 247
Wilhelm Bauer 132
Wilhelm von Humboldt 75
Winston Churchill 127
WIPO 185
Wolfgang Henseler 197
Wolfgang Mommsen 220
Work 2.0 132

Yellow Magic Orchestra 188
Yochai Benkler 228
Yochai Benkler 58
YouTube 201, 207
ZDF 99

Zero-E-Mail Company 129
Ziggie Moondust 189